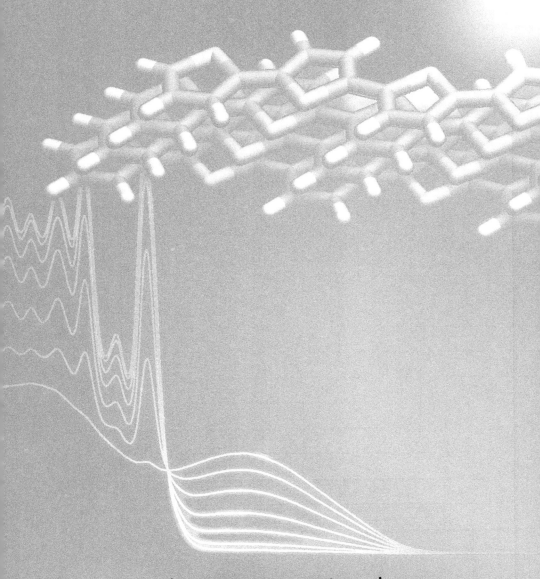

Light-Active Functional Organic Materials

Light-Active Functional Organic Materials

edited by
Hiroko Yamada
Shiki Yagai

Jenny Stanford
Publishing

Published by

Jenny Stanford Publishing Pte. Ltd.
Level 34, Centennial Tower
3 Temasek Avenue
Singapore 039190

Email: editorial@jennystanford.com
Web: www.jennystanford.com

British Library Cataloguing-in-Publication Data
A catalogue record for this book is available from the British Library.

Light-Active Functional Organic Materials

Copyright © 2019 by Jenny Stanford Publishing Pte. Ltd.

All rights reserved. This book, or parts thereof, may not be reproduced in any form or by any means, electronic or mechanical, including photocopying, recording or any information storage and retrieval system now known or to be invented, without written permission from the publisher.

For photocopying of material in this volume, please pay a copying fee through the Copyright Clearance Center, Inc., 222 Rosewood Drive, Danvers, MA 01923, USA. In this case permission to photocopy is not required from the publisher.

ISBN 978-981-4800-15-0 (Hardcover)
ISBN 978-0-429-44853-9 (eBook)

Contents

Preface   xi

1. **Ultraflexible, Fluidic, Optoelectronically Active Molecular/Polymeric Materials**   1
   Akira Shinohara, Chengjun Pan, and Takashi Nakanishi
   - 1.1 Introduction   1
   - 1.2 Background   2
     - 1.2.1 Molecular Design for FMLs via Side Chain Engineering   3
       - 1.2.1.1 Oligo(ethylene glycol)s   3
       - 1.2.1.2 Poly(dimethylsiloxane)   4
       - 1.2.1.3 Linear and branched aliphatic hydrocarbon   5
   - 1.3 Small π-System   7
     - 1.3.1 Luminescent FMLs Based on Anthracene and Pyrene   7
     - 1.3.2 Isothermal Liquefaction in Photoisomerization of Azobenzene Derivatives   11
     - 1.3.3 Implementation of Liquid Carbazole for OLEDs   12
     - 1.3.4 Rod-Shaped π-System   13
     - 1.3.5 Other Small π-Systems   14
   - 1.4 Expanded π-System   15
     - 1.4.1 Porphyrins and Phthalocyanines   15
     - 1.4.2 Carbon Quantum Dots   17
     - 1.4.3 Soft Conjugated Polymers   18
   - 1.5 Globular π-System   19
     - 1.5.1 Permanent Porous Liquid (Cage)   19
     - 1.5.2 Supramolecular FMLs   20
   - 1.6 Summary   21

2. Rigid–Flexible Hybrid Design for Photofunctional Molecules and Materials   27
Shohei Saito

  2.1   Introduction   28
       2.1.1 Hybridization of Rigidity and Flexibility   28
       2.1.2 Controlling Photophysical Properties by Conformation Change   28
       2.1.3 Flapping Fluorophore   30
  2.2   Viscosity Mapping Technique   33
       2.2.1 Molecular Design of Fluorescent Viscosity Probes   33
       2.2.2 Flapping Viscosity Probe   35
             2.2.2.1 Fluorescence and excited-state dynamics   35
             2.2.2.2 Polarity-independent viscochromism   37
             2.2.2.3 Monitoring the epoxy resin curing   38
  2.3   Light-Removable Adhesive   40
       2.3.1 Polymer and Supramolecular Approach   40
       2.3.2 Liquid Crystal Approach   41
       2.3.3 Light-Melt Adhesive   42
             2.3.3.1 Requirements for applications   43
             2.3.3.2 Materials design   44
             2.3.3.3 Adhesive performance   45
             2.3.3.4 Working mechanism   49
  2.4   Conclusion   51

3. Optical Microresonators from π-Conjugated Polymers   57
Yohei Yamamoto

  3.1   Introduction   58
  3.2   Self-Assembled Microspheres from π-Conjugated Polymers   59
  3.3   WGM Photoluminescence from Conjugated Polymer Microspheres   63
  3.4   Inter-sphere Energy Transfer Cascade through Coupled Microspheres   66
  3.5   WGM Lasing from Conjugated Polymer Microspheres   69
  3.6   Summary and Prospects   71

4. **Hydrogen-Bond-Directed Nanostructurization of Oligothiophene Semiconductors for Organic Photovoltaics** — 75

   Xu Lin, Hayato Ouchi, and Shiki Yagai

   4.1 Introduction — 75
   4.2 Insulated Semiconducting Supramolecular Nanorods Composed of Hydrogen-Bonded Oligothiophene Rosettes — 77
   4.3 Semiconducting Supramolecular Nanorods Composed of Hydrogen-Bonded Oligothiophene Rosettes — 79
   4.4 Effect of Substituted Position of Alkyl Chains on the Formation of Semiconducting Supramolecular Nanorods — 80
   4.5 Effect of π-Conjugated Length on the Formation of Semiconducting Supramolecular Nanorods — 83
   4.6 Functionalization of Semiconducting Supramolecular Nanorods — 85
   4.7 Effect of Alkyl Side Chain Length on the Formation and Properties of Supramolecular Nanorods — 86
   4.8 Conclusion — 88

5. **Fundamental Optoelectronic Process in Polymer:Fullerene Heterojunctions** — 93

   Akinori Saeki

   5.1 Introduction — 94
   5.2 Background of Microwave Spectroscopy — 95
   5.3 Charge Separation and Recombination at the Interface of Polymer/Fullerene Bilayer — 98
   5.4 Charge Transport in Polymer:Fullerene Bulk Heterojunction — 102
   5.5 Summary — 108

6. **Near-Infrared Dyes Based on Ring-Expanded Porphyrins with No *Meso*-Bridges** — 115

   Tetsuo Okujima

   6.1 Introduction — 115
   6.2 π-Expanded Cyclo[8]pyrroles — 117

6.3		Dichloride, Carboxylate, and Polyoxometalate Salts	122
6.4		Smaller Analogues	124
6.5		Larger Analogues	125
6.6		Summary	128

7. Gold Isocyanide Complexes Exhibiting Luminescent Mechanochromism and Phase Transitions — 133

Tomohiro Seki and Hajime Ito

7.1		Introduction	133
7.2		Emission Control of Aryl Gold Isocyanide Complexes	135
	7.2.1	Aryl Gold Isocyanide Complexes	135
	7.2.2	Substituents Effect	137
7.3		IR Emissive Mechanochromic Compounds	139
	7.3.1	Extension of π-Conjugated Systems	139
	7.3.2	Anthryl Gold Isocyanide Complexes	140
7.4		Summary	144

8. Thermally Activated Delayed Fluorescence Materials for Organic Light-Emitting Devices — 151

Kyohei Matsuo, Naoya Aizawa, and Takuma Yasuda

8.1		Introduction	151
8.2		Basic Principles of TADF-OLEDs	152
	8.2.1	EL Mechanism in OLEDs	152
	8.2.2	Basic Principles for TADF	154
8.3		Molecular Design for Efficient TADF Molecules	156
	8.3.1	Intramolecular Donor–Acceptor Systems	156
	8.3.2	Intermolecular Donor–Acceptor Systems (Exciplex TADF)	162
8.4		Design for Functionalized TADF Molecules	163
	8.4.1	TADF Molecules with Aggregation-Induced Emission Property	163
	8.4.2	TADF Molecules with Stimuli-Responsive Characteristics	165
	8.4.3	TADF Molecules with Circularly Polarized Luminescence	167
8.5		Summary	168

9. **α-Diketone-Type Precursors of Acenes for Solution-Processed Organic Solar Cells** — 173
   Mitsuharu Suzuki and Hiroko Yamada
   - 9.1 Introduction — 173
   - 9.2 Synthesis of Acenes from α-Diketone-Type Photoprecursors — 175
   - 9.3 Preparation of Acene Thin Films by in situ Conversion of α-Diketone-Type Photoprecursors — 176
     - 9.3.1 Photoprecursor Approach for Solution Processing of Insoluble Acenes — 177
     - 9.3.2 Photoprecursor Approach as an Alternative Means for Processing Soluble Acenes — 179
   - 9.4 Preparation of Organic Photovoltaic Layers via the Photoprecursor Approach — 181
     - 9.4.1 Preparation of BHJ Layers — 181
     - 9.4.2 Application to Layer-by-Layer Solution Deposition — 187
   - 9.5 Summary and Outlook — 190

10. **Photochemical Synthesis of Phenacenes and Their Application to Organic Semiconductors** — 195
    Hideki Okamoto and Yoshihiro Kubozono
    - 10.1 Introduction — 195
      - 10.1.1 General Remarks on Significance of Polycyclic Aromatic Hydrocarbons (PAHs) in Material Sciences — 195
      - 10.1.2 Chemical Structures of PAHs: Acenes and Phenacenes — 197
    - 10.2 Physical Properties of Phenacenes — 198
      - 10.2.1 Stability of Acene and Phenacene — 198
      - 10.2.2 Electronic Features of Acenes and Phenacenes — 198
      - 10.2.3 Solid-State Structure of Phenacenes — 200
    - 10.3 Photochemical Synthesis of Phenacene Frameworks — 202
      - 10.3.1 Conventional Systematic Synthetic Methods of Phenacenes — 202

		10.3.2	Synthesis of Large Phenacenes by Mallory Photocyclization	205
	10.4		OFETs by Using Phenacenes as Active Layer	209
	10.5		Concluding Remark	215

Index   219

Preface

Evolution of molecules that can interact with light may have been a strategy for living organisms to propagate in the early stages of earth with abundant sunlight. In photosynthetic organisms, organic pigments absorb visible light and the resulting excitation energy is efficiently transferred to the reaction center through breathtaking multicomponent molecular systems. Inspired by this epitome of naturally occurring light-active molecular systems, scientists have long devoted their effort to understand how light and molecules interact. Based on a raft of knowledge on light absorption, energy migration and energy/electron transfer, fluorescence and phosphorescence, and various photochemical reactions, etc., we can now utilize light for energy conversion, information storage, medical application, and for the production of new organic materials that cannot be obtained via conventional organic synthesis. For the development of next-generation photoactive molecular materials, however, it is necessary to inclusively overview the state-of-the-art research, spanning sophisticated molecular design, control over the state of molecules, and unconventional analytical methods for these complex materials.

This book is intended to provide a comprehensive overview of some of the cutting-edge π-conjugated molecular and polymer materials for photovoltaics, artificial photosynthesis, and light-emitting devices. These organic materials are created based on unique molecular design and/or self-assembly/self-organization strategies of the contributing authors of this book. Their chapters provide latest insights into the interaction between light and molecules, disclosed by state-of-the-art analytical methods.

The first five chapters focus mainly on the soft materials that have been devised by controlling the assembled state of π-conjugated molecules and macromolecules. In the first chapter, A. Shinohara, C. Pan, and T. Nakanishi describe the design, property, and application of emerging π-conjugated molecular/polymer materials with liquid-like features. Next, S. Saito sheds light on a rigid–flexible hybrid molecular design to produce photofunctional materials in Chapter 2. Y. Yamamoto focuses Chapter 3 on the application of π-conjugated polymer microstructures to optical microresonator. In Chapter 4, X. Lin, H. Ouchi, and S. Yagai overview the supramolecular engineering of small molecular semiconductors for application in organic photovoltaics. In Chapter 5, A. Saeki discusses a special analytical method to observe the optoelectronic process in polymer/fullerene composites.

The last five chapters feature photoactive materials that have been designed focusing more on individual molecular properties. Accordingly, these materials are supposed to be applied in single molecular state or crystalline state to make the best of their intrinsic properties. In Chapter 6, T. Okujima describes near-infrared (NIR) absorbing molecular materials based on a ring-expanded porphyrin design. In Chapter 7, T. Seki and H. Ito overview mechanochromic luminescent crystalline materials based on gold isocyanide complexes. In Chapter 8, K. Matsuo, N. Aizawa, and T. Yasuda shed light on the design of thermally activated delayed fluorescence (TADF) materials for organic light-emitting devices. In Chapter 9, M. Suzuki and H. Yamada review the development of photoprecursor molecular design for the morphology control of the active layer for organic photovoltaics. Finally, in Chapter 10, H. Okamoto and Y. Kubozono describe a photochemical synthetic methodology for phenacene-based crystalline semiconductors.

As the editors of this book, we are deeply indebted to all people who have made the publication of this book successful. First and foremost, we cordially thank all the chapter authors for devoting their time to prepare such high-level manuscripts. Second, we express our gratitude to the team at Pan Stanford Publishing Pte. Ltd., especially Stanford Chong and Jenny Rompas, the directors and publishers, for giving us the opportunity to organize this book, and Shivani Sharma for her help in the editorial process. Finally, we are grateful to all the readers of this book.

Hiroko Yamada
Shiki Yagai
Spring 2019

Chapter 1

Ultraflexible, Fluidic, Optoelectronically Active Molecular/Polymeric Materials

Akira Shinohara,[a] Chengjun Pan,[a] and Takashi Nakanishi[a,b]

[a]*College of Materials Science and Engineering, Shenzhen University, Nanhai Ave 3688, Shenzhen, Guangdong 518060, China*
[b]*International Center for Materials Nanoarchitectonics (WPI-MANA), National Institute for Materials Science (NIMS), 1-1 Namiki, Tsukuba 305-0044, Japan*
nakanishi.takashi@nims.go.jp

1.1 Introduction

Optoelectronically active organic molecules or polymers composed of π-conjugated backbones have extremely rich and tailorable optoelectronic properties such as light harvesting, photoluminescence, and photosensitization. These materials were originally aimed at generation of low-cost, resource-saving, lightweight optoelectronic devices such as light-emitting diodes, solar cells, and field-effect transistor (FET) circuits. In the past decade, an enormous number of chemicals have emerged with

---

Light-Active Functional Organic Materials
Edited by Hiroko Yamada and Shiki Yagai
Copyright © 2019 Jenny Stanford Publishing Pte. Ltd.
ISBN 978-981-4800-15-0 (Hardcover), 978-0-429-44853-9 (eBook)
www.jennystanford.com

high quantum yield, full color, and high stability under ambient atmosphere, light, and heat, which are comparable to conventional inorganic materials. At the same time, another goal of this field is the realization of stretchable and foldable devices. However, this class of materials is intrinsically crystalline and prone to aggregation due to intra- or intermolecular π–π interactions. Consequently, most of the organic optoelectronically active materials are handled as solid or crystal. The lack of solubility and flexibility of these molecules is regarded as a shortcoming in practical application, such as inexpensive processing or developing stretchable/foldable devices. To achieve suitability for those applications, flexible, deformable, or fluidic molecular/polymeric materials are required. As an advantage, these soft materials with nonvolatile nature are also promising for green processing.

Recently, a new generation of solvent-free fluid materials, termed functional molecular liquids (FMLs), has emerged. These materials have made a significant impact on various fields of chemistry, physics, electronics, and life and material sciences [1]. Nowadays, FMLs are classified into a different category: conventional soft materials. In this chapter, we highlight the design, recent developments, and implementation of optoelectronically active FMLs with typical examples divided into small, expanded, and globular π-systems.

1.2 Background

To soften the intrinsically rigid and crystalline π-conjugated optoelectronic functional materials, disturbing intra- or intermolecular π–π stackings is the most essential methodology. Liquefaction strategies are roughly classified into two approaches by using weak ionic interaction like forming ionic liquids and covalent substitution by disturbing substituents. The latter strategy includes low melting temperature (T_m) or glass transition temperature (T_g) contributed by long or branched alkyl, oligo(ethylene glycol) (OEG), and poly(dimethylsiloxane) (PDMS) side chains.

1.2.1 Molecular Design for FMLs via Side Chain Engineering

1.2.1.1 Oligo(ethylene glycol)s

OEGs refer to an oligomer of ethylene oxide composed of –CH$_2$–CH$_2$–O– repeating unit. OEGs are known as low-melting compounds. Replacement of carbon atoms with oxygen atoms in the linear aliphatic hydrocarbon drastically decreases the melting point. For example, octa(ethylene glycol) [HO(CH$_2$CH$_2$O)$_8$H] behaves as a liquid at room temperature (m.p. = 22 °C), while its alkanediol analogue 1,24-tetracosanediol (HO(CH$_2$)$_{24}$OH, 24 atoms in the main chain) shows much higher melting point (108 °C). Low melting point of OEGs is attributed to the lone pairs of electrons on the oxygen atoms of the ethylene glycol repeating unit, which provides a repulsive force and inhibits crystallization of OEG chains. Due to the low-melting feature of OEGs, one can effectively decrease the melting point of intrinsically crystalline π-conjugated molecules.

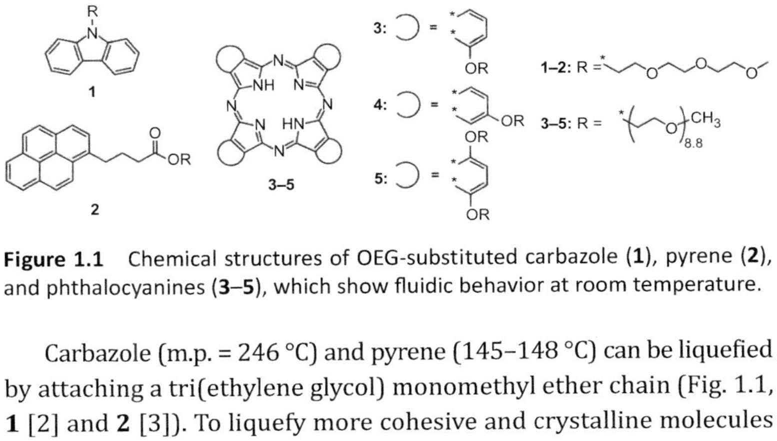

Figure 1.1 Chemical structures of OEG-substituted carbazole (**1**), pyrene (**2**), and phthalocyanines (**3–5**), which show fluidic behavior at room temperature.

Carbazole (m.p. = 246 °C) and pyrene (145–148 °C) can be liquefied by attaching a tri(ethylene glycol) monomethyl ether chain (Fig. 1.1, **1** [2] and **2** [3]). To liquefy more cohesive and crystalline molecules that have expanded π-conjugated systems, it is expected that longer or multiple OEG substituents are required. Phthalocyanine (Pc) is a macrocyclic molecule with highly planer expanded π-plane and does not show melting point up to the decomposition temperature under ambient pressure. Snow et al. reported liquid Pcs **3–5** at room temperature, whose peripheries are substituted by OEG (mean number of repeating units = 8.8) chains at their α- and β-positions [4]. In Fig. 1.1, **3–5** display isotropic transitions close to 10 °C, which would indicate that the OEG substituent dominates this property.

1.2.1.2 Poly(dimethylsiloxane)

PDMS and its derivatives can also be used as effective side chains in FMLs due to not only its low-melting feature but also its high chemical and thermal stability. The binding energy of the Si–O (466 kJ/mol) bond is rather higher than those of C–C (346 kJ/mol) and C–O (358 kJ/mol). PDMS is the most widely used silicon-based organic polymer and is particularly known for its unusual fluidic properties. PDMS behaves as watery fluid at room temperature with a viscosity of 0.1 Pa·s (comparable to olive oil); even its molecular weight reaches 10,000 Da (over hundred –Si–O– repeating units). Compared to the bond length of C–C (1.54 Å) or C–O (1.43 Å), the Si–O (1.64 Å) bond is rather longer; thus, siloxanes would have low barriers for rotation about the Si–O bonds as a consequence of low steric hindrance. By attaching PDMS chains, highly crystalline π-conjugated molecules can be liquefied effectively.

Maya et al. have reported a series of PDMS-substituted Pcs and their lead(II) complexes **6–9** (Fig. 1.2a) [5]. All of the silicone lead Pcs, **6–8**, are isotropic viscous green liquids at room temperature with very low glass transition at –122, –121, and –46 °C, respectively. Metal-free Pc **9** was obtained as blue liquid (T_g = –115 °C) by post-synthetic treatment of **6** with trifluoroacetic acid. PDMS also enables liquefaction of extremely aggregated carbon allotropes. Giannelis et al. have reported multi-walled carbon nanotubes (MWCNTs) that exhibit fluidic behavior in the absence of a diluent or solvent. The solvent-free carbon nanotube (CNT) fluids are synthesized via the oxidation of CNTs by nitric acid following the attachment of a corona of flexible PDMS (molecular weight = 5,000 Da) chains onto the CNT surface [6]. The CNT functionalized with PDMS (**10**) is a homogeneous and viscous tar-like liquid at room temperature (Fig. 1.2b) and dispersible at high concentrations (10 mg/mL) in toluene, providing clear black organosols. Despite its high content of CNTs (~85 wt%), the material maintains a liquid state at room temperature due to the very low T_g of the PDMS chains.

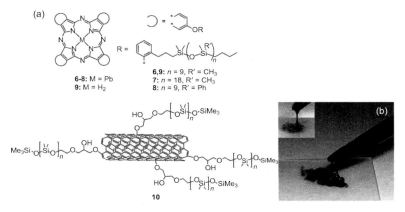

Figure 1.2 (a) Chemical structure of PMDS-substituted liquid phthalocyanines (**6–9**) and multi-walled carbon nanotube (MWCNT) (**10**). (b) Photograph of **10** at room temperature. Reproduced from Ref. [6], Copyright 2007, with permission from Elsevier.

1.2.1.3 Linear and branched aliphatic hydrocarbon

Aliphatic side chains in FMLs can be classified into linear and branched chains. Linear alkyl chains have been used to soften and improve the solubility of π-conjugated molecules for a long time. On the other hand, branched aliphatic alcohols with the chemical formula $HO-CH_2-CH((CH_2)_{n-1}CH_3)-(CH_2)_{n+1}-CH_3$, often abbreviated to $C_nC_{n+4}-OH$, known as Guerbet alcohol [7], are recently used as low-melting substituent. The branching at 2-position and resulting racemic mixtures inhibit their crystallization due to packing disruption. In fact, 24-carbon, 2-decyltetradecanol ($C_{10}C_{14}-OH$) still exists as a liquid at room temperature, while linear aliphatic alcohol becomes solid over dodecyl (C_{12}) chains.

It is astonishing that extremely aggregative fullerenes can be liquefied by attaching aliphatic chains at room temperature (Fig. 1.3a). Solvent-free room temperature liquid C_{60} derivatives **12**, **13**, and **15** were discovered serendipitously by attaching the

2,4,6-tris(alkyloxy)phenyl substituent by our group (Fig. 1.3b) [8a]. Detailed structural studies on the C_{60} and C_{70} fulleropyrrolidine derivatives were also reported [8b]. When the length of the alkyl chain increased, the melting points dramatically decreased from 147–148 °C to −36.5 °C for **11** to **13** (Fig. 1.3c). However, when the alkyl chain was further elongated, the melting points increased. The melting points of **14**, **15**, and **16** were −7.9, 4.5, and 21.5 °C, respectively. The C_{70} derivatives displayed phase behaviors similar to those of the C_{60} derivatives. When the 2,4,6-tris(alkyloxy)phenyl substituent attached to the C_{70} unit, the melting (glassy transition) point gradually increased from −10.5 to 20.6 °C for **20** to **22**. It is noted that unlike C_{60} systems, C_{70} derivatives can be obtained as isomeric mixtures due to the asymmetric structure of their core. This is one of the reasons for melting points similar to those of the corresponding C_{60} derivatives despite the extended π-system of the C_{70} unit. Our group has also synthesized liquid C_{60} derivatives **17**–**19** attached by either Guerbet-alcohol-based alkyl chains (**17**, **18**) or hyperbranched *iso*-octadecyl chains (**19**) [9]. Compared with the linear alkyl-chain-substituted C_{60} derivatives **12**–**16**, which required a 2,4,6-substitution pattern to reach room temperature liquid state, both **17** and **18** needed only two branched chains to generate room temperature liquids. More significantly, **17** and **18** exhibited not only lower melting points (<−120 °C) but also lower viscosity than **12**–**16**. Interestingly, the viscosity of **17** (260 Pa·s) turned out to be much lower than that of **18** (1500 Pa·s), even though **17** had shorter branched alkyl chains. This finding emphasized the importance of substitution at the 2-position on the phenyl unit to produce less viscous C_{60} liquids. However, a further increase in the branching extent (**19**) would lead to both an increased melting point (12 °C) and an enhanced viscosity (128,000 Pa·s) of the isotropic phase, which is even higher than the linear-chain-substituted C_{60} liquids **12**–**16**. This is due to the hyperbranched-structure-induced high intra- and intermolecular friction during flowing. Therefore, both the melting point and viscosity of alkylated C_{60} derivatives can be effectively reduced by suitable branching and proper substitution position of the alkyl chains.

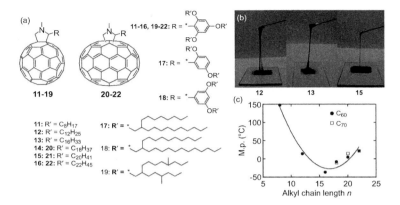

Figure 1.3 (a) Structure of C$_{60}$ (**11–19**) and C$_{70}$ (**20–22**) derivatives. (b) Photographs of room temperature liquid fullerenes **12**, **13**, and **15**. Adapted with permission from Ref. [8a], Copyright 2006, American Chemical Society. (c) Relationship between the melting points and alkyl chain lengths of the fullerene. Adapted with permission from Ref. [8b], Copyright 2013, American Chemical Society.

1.3 Small π-System

1.3.1 Luminescent FMLs Based on Anthracene and Pyrene

Aromatic hydrocarbons, namely arenes, are a class of π-conjugated molecules composed of one or more fused aromatic rings. Arenes (e.g., benzene, naphthalene, anthracene) are considered the simplest functional π-conjugated chromophores and serve as widely utilized substances due to their fundamental interests as well as a wide range of applications. Our group synthesized room temperature FMLs based on anthracene (**23**, **24** in Fig. 1.4) by attaching Guerbet-alcohol-based branched aliphatic chains [10]. The anthracene derivatives **23** and **24** are yellowish transparent viscous fluids at room temperature, whereas **25** is a pasty solid. In the cooling trace of the differential scanning calorimetry (DSC) thermogram, **23** and **24** exhibit a T_g of −59.6 °C and −32.4 °C, respectively. In rheology experiments, the loss modulus G'' is found to be higher than the storage modulus G' over the measured angular frequency range, indicating liquid behavior. Upon doping various luminescent molecules into **23**, its luminescent color can be freely tuned based on Förster resonance energy transfer (FRET) mechanism (Fig. 1.4d).

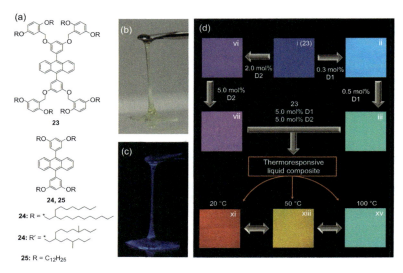

Figure 1.4 (a) Chemical structures of anthracene derivatives **23–25**. Photographs of **23** under (b) visible and (c) UV light (365 nm). (d) Photographs of luminescent tunability and thermal response of the composite of **23**, 9,10-bis(phenylethynyl)anthracene (**D1**) and tris(1,3-diphenyl-1,3-propanedionato)(1,10-phenanthroline)europium(III) (**D2**). Reproduced and modified from Ref. [10] under a Creative Commons Attribution 3.0 Unported (CCBY) license.

Adachi's group proposed unique organic light-emitting diode (OLED) systems using "refreshable" liquid pyrene emitting layer [3]. Although the decomposition of the emitter under the derived current decreases the device luminescence, a quick recovery of emission was observed after the replacement of the decomposed molecules with fresh emitter material through the mesh cathode. The device lifetimes of liquid OLEDs can also be improved by an appropriate combination of guest (**26** in Fig. 1.5a) and host molecules [2b].

Hollamby et al. reported a liquid pyrene compound directly substituted by 2-decyltetradecyl ($C_{10}C_{14}$) chain (**27** in Fig. 1.5a). It was also encapsulated in oil-in-water (o/w) type microemulsions by using a nonionic hexaethylene glycol monodecyl ether ($C_{12}E_6$)

surfactant [11]. In its bulk liquid state, **27** shows luminescence dominated by an intense excimer-like emission ($\lambda_{em} \approx 400$–650 nm) and significantly enhanced fluorescence quantum yield (Φ_{FL} = 0.65) versus the solution state (Φ_{FL} = 0.13). On the other hand, the photoluminescence feature of **27**/$C_{12}E_6$/water microemulsions was found to depend on the amount of **27** loaded to the mixture. The increased excimer emission and droplet size with increasing **27**/$C_{12}E_6$ ratio were attributed to the solvophobicity-driven assembly [12].

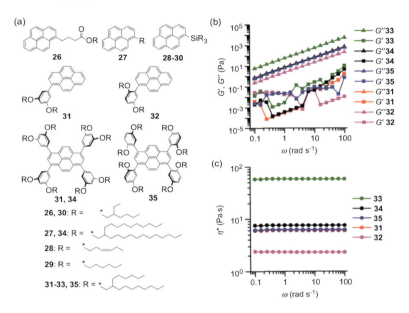

Figure 1.5 (a) Chemical structure of liquid pyrene derivatives. (b) Variation in storage elastic modulus (G', square), viscous loss modulus (G'', triangle), and (c) complex viscosity (η^*, circle) (right) as a function of angular frequency (ω) for neat samples of **31** (red), **32** (pink), **33** (green), **34** (black), and **35** (blue) with a constant strain amplitude γ = 0.25 at 25 °C. Reproduced from Ref. [14] under a Creative Commons Attribution 4.0 Unported (CCBY) license.

Yamaguchi et al. have reported similar o/w fluorescent nanoparticles composed of trialkylsilyl-substituted liquid pyrenes **28–30** (Fig. 1.5a) by using a mixture of phosphatidylcholine (DSPC) and cholesterol as surfactant [13]. While triisopropyl-substituted pyrene is a solid with a melting point at 100 °C, **28–30** are liquid at room temperature, which show low glassy transitions at −71 °C, −58 °C, and −57 °C, respectively. An efficient FRET from **28–30** excimer to spontaneously encapsulated hydrophobic dopant dyes such as green-emissive C545T and red-emissive 2-tert-butyl-4-(dicyanomethylene)-6-[2-(1,1,7,7-tetramethyljulolidin-9-yl)vinyl]-4H-pyran (DCJTB) was observed. A continuous emission color tuning of the nanoparticles was achieved by simply changing conditions in nanoparticles formation (i.e., concentration and combination of dopants).

Very recently, our group synthesized 1-substituted (**31**, **32**) and 1,3,6,8-tetrasubstituted (**33–35**) pyrene derivatives (Fig. 1.5a). They exhibit viscous liquid states, which show glassy transitions from −63.5 to −43.0 °C, while reference methoxy-substituted counterparts are solid with high melting points [14]. Polarized optical microscopy (POM) images of fluids **31–35** at room temperature do not provide any texture, indicating a lack of long-range ordered and assembled domains. The absence of long-range order is further supported by small and wide-angle X-ray scattering (SWAXS), which exhibit only broad halos. The random orientation of pyrene cores are further evidenced by flash-photolysis time-resolved microwave conductivity (FP-TRMC) [15]. According to FP-TRMC profiles, **31–35** show pseudo-photoconductivity maxima of the order of 10^{-5}–10^{-4} cm^2/Vs and a prompt deactivation faster than 0.3 μs. These values are considerably lower than those reported for other crystalline and liquid crystalline compounds [16]. DSC, POM, SWAXS, FTIR, and FP-TRMC analyses all indicate that **31–35** are amorphous fluids with random molecular orientations at room temperature, even though they have different alkyl-chain substituent styles. We also performed rheological study to confirm the fluidic behavior of **31–35** (Figs. 1.5b,c). While dynamic frequency sweep, complex viscosity (η^*) of liquid pyrenes are constant, which indicates Newtonian liquid behavior.

Very recently, a much smaller arene, i.e., naphthalene, has also been liquefied by attaching branched alkyl chains [17]. In the study, 1-alkyloxy regioisomers exhibited excimeric fluorescence, whereas 2-alkyloxy derivatives showed monomer-rich fluorescence features. The position of the alkyl-chain substituent can alter the electronic structure and molecular packing of the naphthalene unit; therefore, the effect of regioisomerism on the photophysical, calorimetric, and rheological properties was well documented.

1.3.2 Isothermal Liquefaction in Photoisomerization of Azobenzene Derivatives

Azobenzenes are one of the most common photochromic molecules that show reversible *trans-/cis*-isomerization driven by UV/visible light irradiation. Photoinduced isomerization from thermally stable *trans*-isomer to metastable *cis*-isomer leads to a drastic change in their mechanical, chemical as well as photochemical properties due to lower symmetry in the latter. These compounds have been employed for photofunctional systems such as microscopic phase transition from liquid crystal to isotropic, sol–gel transition. While thermally reversible phase transition is a commonsense for most chemical compounds, photochemical liquefaction under isothermal conditions has been rather rare.

The photochemically reversible liquefaction–solidification cycles at room temperature had been reported by Noritake et al. [18]. Alkylated dimeric (**36**) and trimeric (**37**) azobenzenophanes show photo-irradiated crystal to liquid phase transitions, while the monomer counterpart does not show such behavior (Fig. 1.6). Detailed crystallographic studies of **36** were performed using bright synchrotron X-ray to elucidate structural features governing the photoinduced crystal–liquid phase transition [19].

Akiyama et al. also reported similar liquefaction–solidification cycles by means of a series of azobenzene-modified sugar alcohols [20]. The *trans*-isomers of **38**–**42** (Fig. 1.6) melt to be their *cis*-form by UV irradiation for up to an hour under isothermal conditions. The obtained fluids gradually turned to solid after 2 days to be their *trans*-form. The reversible liquefaction–solidification property had been successfully applied as photoresponsive adhesives [21].

Our group synthesized an azobenzene liquid via substitution by 2-octyldedecyl (C_8C_{12}) branched alkyl chain **43** (Fig. 1.6) [22] and studied its hydrophobic–amphiphilicity feature in solution state. Kimizuka et al. synthesized similar liquid azobenzene attached with 2-ethylhexyl (C_2C_6) chain **44** (Fig. 1.6) and explored its application for solar thermal fuel under solvent-free liquid conditions [23]. The *trans*-isomer of **44** is a yellowish liquid with a low T_g at −63 °C and stable up to 220 °C without thermal decomposition. The enthalpy (ΔH) and energy density of the *cis*-**44** are calculated to be 52 kJ/mol and 47 Wh/kg, respectively.

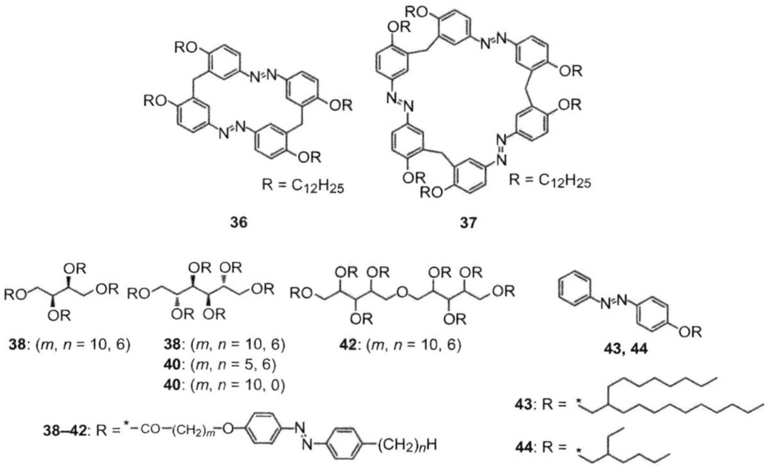

Figure 1.6 Chemical structure of azobenzene derivatives showing isothermal solid/liquid phase transition induced by photoisomerization (**36–42**), possessing hydrophobic amphiphilicity (**43**) and heat storage nature (**44**).

1.3.3 Implementation of Liquid Carbazole for OLEDs

Peyghambarian et al. have reported a room temperature carbazole liquid **45** by attaching a C_2C_6 chain to nitrogen atom (Fig. 1.7), which was employed as a solvent in the ellipsometry measurement for the determination of electric-field-induced birefringence in photorefractive polymer composites [24]. The enhanced charge carrier mobility (4×10^{-6} cm/Vs) of **45** is attributed to both a larger transfer integral and changes in the distribution of the excimer trapping sites [25]. Various liquid OLEDs have been successfully composed by using **45** doped with different guest emitters such as 5,6,11,12-tetraphenylnaphthacnee (rubrene). Adachi et al. proposed that in OLEDs composed of the fluid organic semiconductors **45** can be circulated or refilled, fresh semiconductor materials will always be utilized and which will minimize device degradation [26]. In another report, an OLED composed of **45** as an emitting layer and TiO_2 as a hole-blocking layer was prepared [27].

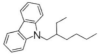

Figure 1.7 Chemical structure of a liquid carbazole derivative **45**.

1.3.4 Rod-Shaped π-System

The linear π-conjugated molecules oligo(*p*-phenylenevinylene)s (OPVs) have been widely studied in organic optoelectronics because they have excellent stability and emission characteristics, as well as self-assembly properties. Our group has synthesized room temperature liquid OPVs by substituting with branched alkyl chains (**46–49** in Fig. 1.8) [28]. These OPVs are pale yellow liquids at room temperature with T_g between −43 °C and −55 °C. Branched alkyl chains substituting on the 2,4,6-positions (**46, 47**) introduce much lower viscosity than those appending on the 3,5-positions (**48, 49**). The UV-vis absorption and fluorescence spectroscopic properties of these neat liquids are almost identical to their dilute solution analogues demonstrating efficient isolation of OPV unit upon enveloping by bulky and flexible branched alkyl chains in the solvent-free state. We have employed the blue-emitting OPVs as matrix for doping green-emitting tris(8-hydroxyquinolinato) aluminum (Alq$_3$) and orange-emitting rubrene to obtain white-emitting liquids. Simply blending the three compounds for just 1 min would result in the composites exhibiting a white emission spanning from 400 to 700 nm. The composite maintained the strong emission features of OPV and had a quantum yield of more than 35%, which, together with the low T_g at around −45 °C and low viscosity of 3.2 Pa·s, enabled writing or painting it on various surfaces, i.e., UV light-emitting diode (LED).

As an analogue to OPVs, Adachi et al. synthesized π-conjugated oligomers of fully planar oligo(*p*-phenyleneethynylene) (OPE) backbone with branched alkyl chains (**50** in Fig. 1.8) [29]. The melting point of **50** is at 20 °C, and it maintains a liquid state above the melting temperature. The UV-vis absorption peaks of the liquid state at 310 and 375 nm were almost identical to the diluted solution in CHCl$_3$. The latter band corresponds to the π–π* transition of the phenyleneethynylene core. By contrast, the emission peak at around 500 nm of the liquid-state **50** shows broader and red-shifted from the maximum in CHCl$_3$. These results suggest the formation of excimer-like structures because of an increase in π–π interactions between phenyleneethynylene units at excited state.

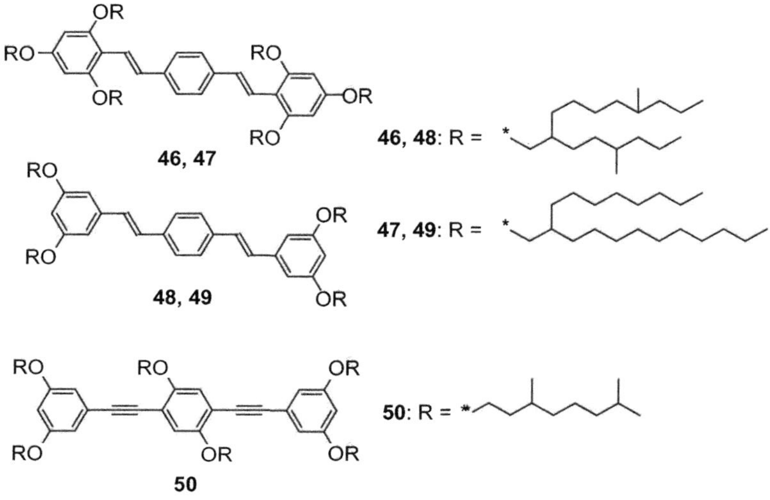

Figure 1.8 Room-temperature FMLs with rod-shaped OPV (**46–49**) and OPE (**50**) core.

1.3.5 Other Small π-Systems

Ishi-i et al. reported a room temperature liquid benzothiadiazole **51** (Fig. 1.9) by attaching the C$_2$C$_6$ chains, as a dark-red viscous liquid [30]. Unlike liquid pyrene derivatives as mentioned earlier, fluorescence quantum yield Φ$_{FL}$ of **51** is significantly lower than in solution. Audebert et al. reported tetrazine-based fluorescent liquid (**52**, **53** in Fig. 1.9) with quite low glass transition temperature (<−60 °C) and low viscosity (28 mPa·s for **52**, 58 mPa·s for **53**) at room temperature [31]. Kokado et al. have reported the liquefaction of tetraphenylethylene (TPE) derivatives **54–57** (Fig. 1.9) by using similar alkylation strategy [32]. Interestingly, solvent-free liquids **54–57** show higher fluorescence efficiency rather than dodecyl-substituted solid or bear TPE.

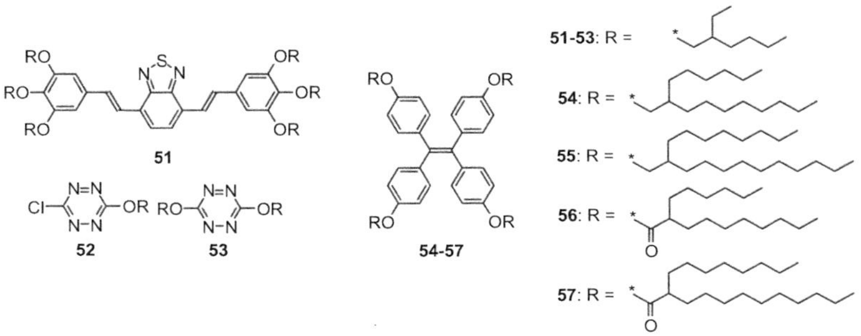

Figure 1.9 Chemical structures of liquid benzothiadiazole (**51**), tetrazines (**52, 53**), and TPEs (**54–57**) that show liquid behavior at room temperature.

1.4 Expanded π-System

1.4.1 Porphyrins and Phthalocyanines

Porphyrins and phthalocyanines are known as highly planar heteronuclear π-conjugated compounds and attract considerable interests due to their characteristic photophysical properties such as unusually large molar absorption coefficient, photoluminescence, photosensitization, and two-photon absorption cross section. The synthesis of room temperature liquid porphyrins by attaching 3,4,5-trialkyloxylphenyl substituents on their *meso*-positions was independently reported by the research groups of Gryko [33a] and Maruyama [33b]. The 5,10,15,20-tetrakis(3,4,5-tri(decyloxy)phenyl)porphyrin **58** and its undecyloxy analogue **59** (Fig. 1.10) are exhibit as liquid at room temperature with low melting temperature at −55 °C and −24 °C, respectively (Fig. 1.10). The liquid behavior was confirmed by rheology analysis. However, a further increase or decrease in the chain length results in a melting point above room temperature. This behavior is very similar to that of liquid C_{60} derivatives [8].

Very recently, a liquid-like porphyrin–metal complex was reported by Hasegawa et al. [34]. Synchrotron microbeam glazing-incidence X-ray scattering (GIXS) of its analogue found characteristic amorphous halo patterns approximately at q of 13.5 nm^{-1} and no periodic spacings were found other than primary intermolecular spacing. They also developed π-conjugated porphyrin glasses based on **60** (Fig. 1.10), which display remarkable Stokes shift toward the near-infrared region through the aid of exciton–photon coupling.

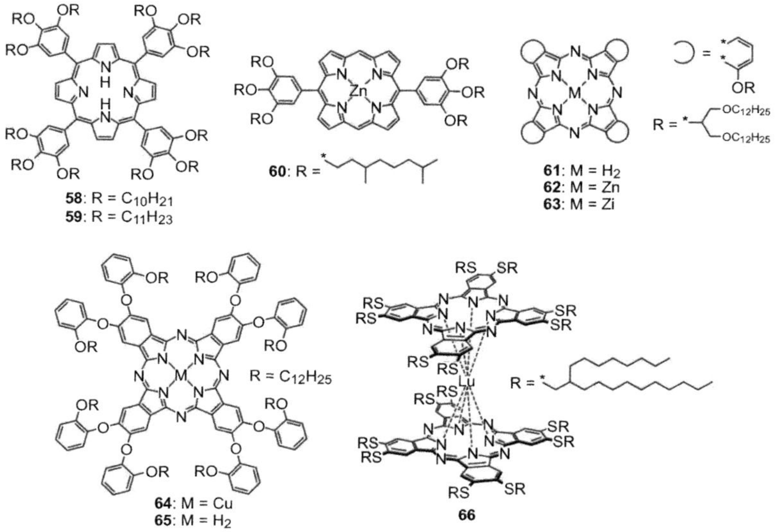

Figure 1.10 Room temperature liquid porphyrins (**58–60**) and phthalocyanines (**61–66**).

Ahsen et al. reported that a metal-free phthalocyanine derivative **61** is fluidic at room temperature as well as the Zn(II) and Ni(II) complexes (**62, 63**) (Fig. 1.10) [35]. Non-periphery, α-substituted phthalocyanines **61–63** are isotropic, while periphery β-substituted phthalocyanines were found as liquid crystalline at room temperature.

Very recently, two other liquid phthalocyanines with linear (**64, 65**) [36] or branched alkyl chain (**66**) (Fig. 1.10) [37] substituents have been reported by our group. Both liquid systems show non-Newtonian liquid behavior due to subtle intermolecular π–π interactions among neighboring phthalocyanine units. In the latter case, a double-decker lutetium phthalocyanine was chosen as the multi- and switchable functional core unit, such as spin-active nature and electrochromic behavior.

1.4.2 Carbon Quantum Dots

Fluorescent carbon nanoparticles or carbon quantum dots (CQDs), which are generally small carbon nanoparticles (less than 10 nm in size), are a new class of carbon nanomaterials that have emerged during the last few years and have attracted much interest as potential competitors to conventional semiconductor quantum dots. Hao et al. reported solvent-free liquid CQDs at room temperature [38]. Pyrolyzation of 1-methyl-3-aminopropylimidazolium bromide ([APMIm]$^+$[Br]$^-$) and citric acid at 180–260 °C for 1–4 h under argon gave water-soluble, ionic liquid unit modified CQDs **67** (Fig. 1.11a) as brown powder. Liquid CQDs were obtained by anion exchange from Br$^-$ to NTf^{2-} (Fig. 1.11b). Rogach et al. have reported flexible ionogel by a combination of a carboxyl-functionalized ionic liquid 1-(3-carboxypropyl)-3-butylimidazolium bromide with organosilane-functionalized carbon dots (**68**) [39]. Ionogel composed of **68** shows full-color emission tunability simply by its thickness explained by consecutive reabsorptions and re-emission. Emission color was turned from blue to red, which covers the whole visible spectral range with increasing thickness of the ionogel from 0.1 mm to 10.0 mm (Fig. 1.11c).

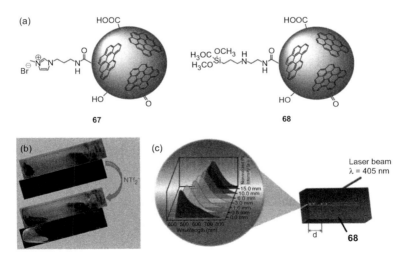

Figure 1.11 (a) Structure of CQDs (**67**, **68**). (b) Anion exchange in **67** results room temperature liquid CQDs. Reproduced with permission from Ref. [38], Copyright 2015, American Chemical Society. (c) Thickness-dependent full-color tuning of emission in ionogel consisting of **68**. Adapted with permission from Ref. [39], Copyright 2014, American Chemical Society.

1.4.3 Soft Conjugated Polymers

π-Conjugated polymers (CPs) have attracted considerable interests for organic optoelectronic applications owing to their distinguishing mechanical properties, processability, and electronic conductivity. In fact, CPs have become key materials in the field of organic optoelectronics. Conventional CPs, such as polythiophene (PT), polyfluorene (PF), polyphenylene (PP), polyphenylenevinylene (PPV), and polyphenyleneethynylene (PPE), have planar geometries and strong intermolecular interactions due to the extremely stiff and rigid main chains fully composed of sequent sp or sp^2 carbons and tendency of π–π stacking, resulting in a highly cofacial chain packing in their solid state. Softening the generally stiff and rigid π-conjugated polymers is a considerable challenge for their practical applications. Kwak et al. reported gum-like conjugated polymer based on poly(diphenylacetylene) (PDPA) derivatives (**69, 70** in Fig. 1.12a), which have very unusual and highly twisted backbone due to steric repulsion between the neighboring bulky phenyl groups [40]. The same group also developed soft CPs sulfonated PDPA **71** as anionic polyelectrolytes by complexation with cationic surfactants (octadecyl)$_x$(methyl)$_y$ ammonium bromides (O_xM_yABs) having different numbers of long hydrophobic alkyl chains [41]. The melting temperatures after complexation are at 12.5 °C, 14.3 °C, and 6.1 °C for **71**-O_1M_3AB, **71**-O_2M_2AB, and **71**-O_3M_1AB, respectively.

Sugiyasu et al. reported thiophene-based unique insulated molecular wires (IMWs) [42]. All of **72–75** are emissive in the entire visible region and show high fluorescence quantum yield (Φ_{FL}) up to 44% even in the film state, while IMWs in the other reports are generally around 10–20% of Φ_{FL} (Fig. 1.12b). After thermal annealing of a drop-cast film of **74** at 80 °C for 30 min, a fluorescent, flexible, self-standing film was obtained. A small piece of the film could sustain a weight of more than 200 g (>5 MPa of tensile stress), was stretchable (>300% of tensile strain), and even foldable without the formation of any cracks.

Compared to small and linear π-systems, liquid-expanded π-systems are still rare due to their high tendency to π–π stacking. Müllen et al. synthesized hexabenzocoronene (HBC) molecule periphery substituted by $C_{10}C_{14}$ chains [43]. Phase transition from a waxy-soft mesophase to the isotropic was still higher than room temperature at 46 °C in HBC–$C_{10}C_{14}$.

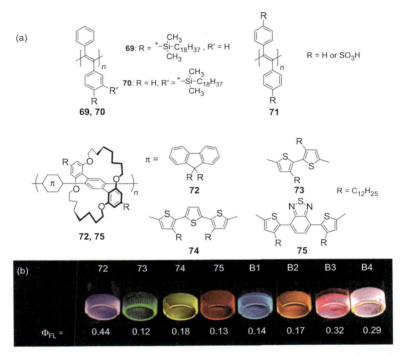

Figure 1.12 (a) Chemical structure of conjugated polymers (**69–75**). (b) Photographs of the films of **72–75** and blend films **B1–B4** taken under UV (365 nm) illumination. The blending ratios **72/73/74/75** of **B1**, **B2**, **B3**, and **B4** are 100/10/0/0, 0/0/100/5, 100/0/0/5, and 200/0/3/1, respectively. Adapted with permission from Ref. [42], Copyright 2013, John Wiley and Sons.

1.5 Globular π-System

1.5.1 Permanent Porous Liquid (Cage)

Apart from the aforementioned optoelectronic functions in FMLs based on a π-conjugated system, geometrical functions such as porosity in FMLs are of alternative interest in the emerging liquid science. James's group reported unique porous liquids composed of rigid organic cage molecules and having permanent porosity, while other common liquids have limited porosity due to their poorly defined and transient intermolecular cavities [44]. The cage molecule **76** has been prepared by taking rigid cage architecture, which defines a molecular pore space, and dissolving them at high concentration (~44 wt%) in 15-crown-5 as solvent, which is too large to enter the pores (Fig. 1.13). The cage molecule has a cavity of ~5 Å diameter at its center, which is accessible through four access windows of ~4 Å in diameter. The viscosity of the mixture liquid was 20 mPa·s to >140 mPa·s in the range of 25–50 °C. Molecular dynamics (MD) and positron (e^+) annihilation lifetime spectroscopy (PALS) experiments suggest that the cages in the porous liquid were empty. The solubility of methane in the porous liquid was 52 μmol/g at 30 °C, while 6.7 μmol/g in pure 15-crown-5.

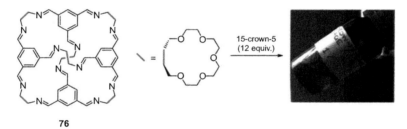

Figure 1.13 Structure of cage molecule **76** and its solution "porous liquid" in 15-crown-5 (**76**:15-crown-5 = 1:12). Reproduced with permission from Ref. [44], Copyright 2015, Springer Nature.

1.5.2 Supramolecular FMLs

In a complexation event in host–guest (H–G) supramolecular systems, the high stability and efficiency of the H–G complex is essential to their practical application. Ogoshi et al. proposed solvent-free, nonvolatile liquid host molecules, named cyclic host liquids (CHLs), to prevent solvation of H and/or G molecules, which act as a barrier for H–G complexation. The CHLs based on pillar[5]arene (**77**, **78**), α-, β-, and γ-cyclodextrins (CDs, **79**–**81**) were obtained by attaching hydrophilic triethylene glycol chains (Fig. 1.14) [45]. Liquid [2]rotaxane **82** was obtained by the threading of aliphatic chain following cationation interlocking under solvent-free conditions in high yield (91%, isolated).

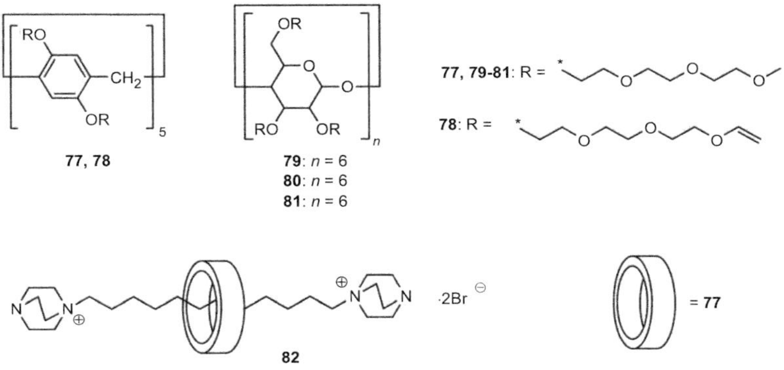

Figure 1.14 The CHLs based on pillar[5]arene (**77**, **78**), α-, β-, and γ-CDs (**79–81**) and liquid [2]rotaxane (**82**).

1.6 Summary

In this chapter, the design, synthesis, and application of various FMLs have been discussed. In the last decade, nonvolatile π-conjugated FMLs have been found useful for optoelectronics. Various strategies for liquefaction have been carried out by employing ionic liquid, bulky branched alkyl chains, poly(ethylene glycol) or poly(dimethylsiloxane) chains on π-chromophore, softening the intrinsically stiff and rigid π-system molecules. It has become possible to improve the lack of processability and of homogeneous fabrication of π-conjugated materials. Despite these fascinating properties, there have been few studies on the development of application of FMLs compared with the dramatic progress in conventional π-conjugated functional molecules. The history of fabricating these FMLs into flexible/stretchable optoelectronic devices is still short at present and further exploration is required. Thus, it can be reasonably expected that more sophisticated device systems with unique functions originated from FMLs would emerge in the near future.

References

1. (a) Babu, S. S. and Nakanishi, T. (2013). Nonvolatile functional molecular liquids, *Chem. Commun.*, **49**, pp. 9373–9382; (b) Ghosh, A. and Nakanishi, T. (2017). Frontiers of solvent-free functional molecular liquids, *Chem. Commun.*, **53**, pp. 10344–10357.
2. (a) Kubota, K., Hirata, S., Shibano, Y., Hirata, O., Yahiro, M., and Adachi, C. (2012). Liquid carbazole substituted with a poly(ethylene oxide) group and its application for liquid organic light-emitting diodes, *Chem. Lett.*, **41**, pp. 934–936; (b) Hirata, S., Heo, H., Shibano, Y., Hirata, O., Yahiro, M., and Adachi, C. (2012). Improved device lifetime of organic light emitting diodes with an electrochemically stable π-conjugated liquid host in the liquid emitting layer, *Jpn. J. Appl. Phys.*, **51**, 041604.
3. Shim, C., Hirata, S., Oshima, J., Edura, T., Hattori, R., and Adachi, C. (2012). Uniform and refreshable liquid electroluminescent device with a back side reservoir, *Appl. Phys. Lett.*, **101**, 113302.
4. Snow, A. W., Shirk, J. S., and Pong, R. G. S. (2000). Oligooxyethylene liquid phthalocyanines, *J. Porphyrins Phthalocyanines*, **4**, pp. 518–524.
5. (a) Maya, E. M., Shirk, J. S., Snow, A. W., and Roberts, G. L. (2001). Peripherally-substituted polydimethylsiloxane phthalocyanines: A novel class of liquid materials, *Chem. Commun.*, pp. 615–616; (b) Maya, E. M., Snow, A. W., Shirk, J. S., Pong, R. G. S., Flom, S. R., and Roberts, G. L. (2003). Synthesis, aggregation behavior and nonlinear absorption properties of lead phthalocyanines substituted with siloxane chains, *J. Mater. Chem.*, **13**, pp. 1603–1613.
6. Bourlinos, A. B., Georgakilas, V., Boukos, N., Dallas, P., Trapalis, C., and Giannelis, E. P. (2007). Silicone-functionalized carbon nanotubes for the production of new carbon-based fluids, *Carbon*, **45**, pp. 1583–1595.
7. O'Lenick Jr., A. J. (2001). Guerbet chemistry, *J. Surfact. Deterg.*, **4**, pp. 311–315.
8. (a) Michinobu, T., Nakanishi, T., Hill, J. P., Funahashi, M., and Ariga, K. (2006). Room-temperature liquid fullerenes: An uncommon morphology of C_{60} derivatives, *J. Am. Chem. Soc.*, **128**, pp. 10384–10385; (b) Michinobu, T., Okoshi, K., Murakami, Y., Shigehara, K., Ariga, K., and Nakanishi, T. (2013). Structural requirements for producing solvent-free room temperature liquid fullerenes, *Langmuir*, **29**, pp. 5337–5344.
9. Li, H., Babu, S. S., Turner, S. T., Neher, D., Hollamby, M. J., Seki, T., Yagai, S., Deguchi, Y., Möhwald, H., and Nakanishi, T. (2013). Alkylated-C_{60}

based soft materials: Regulation of self-assembly and optoelectronic properties by chain branching, *J. Mater. Chem. C*, **1**, pp. 1943–1951.
10. Babu, S. S., Hollamby, M. J., Aimi, J., Ozawa, H., Saeki, A., Seki, S., Kobayashi, K., Hagiwara, K., Yoshizawa, M., Möhwald, H., and Nakanishi, T. (2013). Nonvolatile liquid anthracenes for facile full-colour luminescence tuning at single blue-light excitation, *Nat. Commun.*, **4**, 1969.
11. Hollamby, M. J., Danks, A. E., Schnepp, Z., Rogers, S. E., Hart, S. R., and Nakanishi, T. (2016). Fluorescent liquid pyrene derivative-in-water microemulsions, *Chem. Commun.*, **52**, pp. 7344–7347.
12. Maggini, L. and Bonifazi, D. (2012). Hierarchised luminescent organic architectures: Design, synthesis, self-assembly, self-organization and functions, *Chem. Soc. Rev.*, **41**, pp. 211–241.
13. Taki, M., Azeyanagi, S., Hayashi, K., and Yamaguchi, S. (2017). Color-tunable fluorescent nanoparticles encapsulating trialkylsilyl-substituted pyrene liquids, *J. Mater. Chem. C*, **5**, pp. 2142–2148.
14. Lu, F., Takaya, T., Iwata, K., Kawamura, I., Saeki, A., Ishii, M., Nagura, K., and Nakanishi, T. (2017). A guide to design functional molecular liquids with tailorable properties using pyrene-fluorescence as a probe, *Sci. Rep.*, **7**, 3416.
15. Saeki, A., Koizumi, Y., Aida, T., and Seki, S. (2012). Comprehensive approach to intrinsic charge carrier mobility in conjugated organic molecules, macromolecules, and supramolecular architectures, *Acc. Chem. Res.*, **45**, pp. 1193–1202.
16. (a) Saeki, A., Seki, S., Takenobu, T., Iwasa, Y., and Tagawa, S. (2008). Mobility and dynamics of charge carriers in rubrene single crystals studied by flash-photolysis microwave conductivity and optical spectroscopy, *Adv. Mater.*, **20**, pp. 920–923; (b) Xiao, Q., Sakurai, T., Fukino, T., Akaike, K., Honsho, Y., Saeki, A., Seki, S., Kato, K., Takata, M., and Aida, T. (2013). Propeller-shaped fused oligothiophenes: A remarkable effect of the topology of sulfur atoms on columnar stacking, *J. Am. Chem. Soc.*, **135**, pp. 18268–18271.
17. Narayan, B., Nagura, K., Takaya, T., Iwata, K., Shinohara, A., Shinmori, H., Wang, H., Li, Q., Sun, X., Li, H., Ishihara, S., and Nakanishi, T. (2018). The effect of regioisomerism on the photophysical properties of alkylated-naphthalene liquids, *Phys. Chem. Chem. Phys.*, **20**, pp. 2970–2975.
18. (a) Norikane, Y., Hirai, Y., and Yoshida, M. (2011). Photoinduced isothermal phase transitions of liquid-crystalline macrocyclic azobenzenes, *Chem. Commun.*, **47**, pp. 1770–1772; (b) Uchida, E., Sasaki, K., Nakamura, Y., Azumi, R., Hirai, Y., Akiyama, H., Yoshida, M., and Noritake, Y. (2013). Control of the orientation and photoinduced

phase transitions of macrocyclic azobenzene, *Chem. Eur. J.*, **19**, pp. 17391–17397.

19. Hoshino, M., Uchida, E., Norikane, Y., Azumi, R., Nozawa, S., Tomita, A., Sato, T., Adachi, S., and Koshihara, S. (2014). Crystal melting by light: X-ray crystal structure analysis of an azo crystal showing photoinduced crystal-melt transition, *J. Am. Chem. Soc.*, **136**, pp. 9158–9164.

20. Akiyama, H. and Yoshida, M. (2012). Photochemically reversible liquefaction and solidification of single compounds based on a sugar alcohol scaffold with multi azo-arms, *Adv. Mater.*, **24**, pp. 2353–2356.

21. Akiyama, H., Kanazawa, S., Okuyama, Y., Yoshida, M., Kihara, H., Nagai, H., Norikane, Y., and Azumi, R. (2014). Photochemically reversible liquefaction and solidification of multiazobenzene sugar-alcohol derivatives and application to reworkable adhesives, *ACS Appl. Mater. Interfaces*, **6**, pp. 7933–7941.

22. Hollamby, M. J., Karny, M., Bomans, P. H. H., Sommerdjik, N. A. J. M., Saeki, A., Seki, S., Minamikawa, H., Grillo, I., Pauw, B. R., Brown, P., Eastoe, J., Möhwald, H., and Nakanishi, T. (2014). Directed assembly of optoelectronically active alkyl–π-conjugated molecules by adding *n*-alkanes or π-conjugated species, *Nat. Chem.*, **6**, pp. 690–696.

23. Masutani, K., Morikawa, M., and Kimizuka, N. (2014). A liquid azobenzene derivative as a solvent-free solar thermal fuel, *Chem. Commun.*, **50**, pp. 15803–15806.

24. Hendrickx, E., Guenther, B. D., Zhang, Y., Wang, J. F., Staub, K., Zhang, Q., Marder, S. R., Kippelen, B., and Peyghambarian, N. (1999). Ellipsometric determination of the electric-field-induced birefringence of photorefractive dyes in a liquid carbazole derivative, *Chem. Phys.*, **245**, pp. 407–415.

25. Ribierre, J. -C., Aoyama, T., Muto, T., Imase, Y., and Wada, T. (2008). Charge transport properties in liquid carbazole, *Org. Electron.*, **9**, pp. 396–400.

26. Xu, D. and Adachi, C. (2009). Organic light-emitting diode with liquid emitting layer, *Appl. Phys. Lett.*, **95**, 053304.

27. Hirata, S., Kubota, K., Jung, H., Hirata, O., Goushi, K., Yahiro, M., and Adachi, C. (2011). Improvement of electroluminescence performance of organic light-emitting diodes with a liquid-emitting layer by introduction of electrolyte and a hole-blocking layer, *Adv. Mater.*, **23**, pp. 889–893.

28. Babu, S. S., Aimi, J., Ozawa, H., Shirahata, N., Saeki, A., Seki, S., Ajayaghosh, A., Möhwald, H., and Nakanishi, T. (2012). Solvent-free luminescent organic liquids, *Angew. Chem. Int. Ed.*, **51**, pp. 3391–3395.

29. Adachi, N., Itagaki, R., Sugeno, M., and Norioka, T. (2014). Dispersion of fullerene in neat synthesized liquid-state oligo(*p*-phenyleneethynylene)s, *Chem. Lett.*, **43**, pp. 1770–1772.

30. Ishi-i, T., Sakai, M., and Shinoda, C. (2013). Benzothiadiazole-based dyes that emit red light in solution, solid, and liquid state, *Tetrahedron*, **69**, pp. 9475–9480.

31. Allain, C., Piard, J., Brosseau, A., Han, M., Paquier, J., Marchandier, T., Lequeux, M., Boissière, C., and Audebert, P. (2016). Fluorescent and electroactive low-viscosity tetrazine-based organic liquids, *ACS Appl. Mater. Interfaces*, **8**, pp. 19843–19846.

32. Machida, T., Taniguchi, R., Oura, T., Sada, K., and Kokado, K. (2017). Liquefaction-induced emission enhancement of tetraphenylethene derivatives, *Chem. Commun.*, **53**, pp. 2378–2381.

33. (a) Nowak-Król, A., Gryko, D., and Gryko, D. T. (2010). Meso-substituted liquid porphyrins, *Chem. Asian. J.*, **5**, pp. 904–909; (b) Maruyama, S., Sato, K., and Iwahashi, H. (2010). Room temperature liquid porphyrins, *Chem. Lett.*, **39**, pp. 714–716.

34. Morisue, M., Ueno, I., Nakanishi, T., Matsui, T., Sasaki, S., Shimizu, M., Matsui, J., and Hasegawa, Y. (2017). Amorphous porphyrin glasses exhibit near-infrared excimer luminescence, *RSC Adv.*, **7**, pp. 22679–22683.

35. Durmuş, M., Lebrun, C., and Ahsen, V. (2004). Synthesis and characterization of novel liquid and liquid crystalline phthalocyanines, *J. Porphyrins Phthalocyanines*, **8**, pp. 1175–1186.

36. Chino, Y., Ghosh, A., Nakanishi, T., Kobayashi, N., Ohta, K., and Kimura, M. (2017). Stimuli-responsive viscoelastic properties for liquid phthalocyanines, *Chem. Lett.*, **46**, pp. 1539–1541.

37. Zielinska, A., Takai, A., Sakurai, H., Saeki, A., Leonowicz, M., and Nakanishi, T. (2018). A spin-active, electrochromic, solvent-free molecular liquid based on double-decker lutetium phthalocyanine bearing long branched alkyl chains, *Chem. Asian J.*, 13, pp. 770–774

38. Wang, B., Song, A., Feng, L., Ruan, H., Li, H., Dong, S., and Hao, J. (2015). Tunable amphiphilicity and multifunctional applications of ionic-liquid-modified carbon quantum dots, *ACS Appl. Mater. Interfaces*, **7**, pp. 6919–6925.

39. Wang, Y., Kalytchuk, S., Zhang, Y., Shi, H., Kershaw, S. V., and Rogach, A. L. (2014). Thickness-dependent full-color emission tunability in a flexible carbon dot ionogel, *J. Phys. Chem. Lett.*, **5**, pp. 1412–1420.

40. Jin, Y., Bae, J., Cho, K., Lee, W., Hwang, D., and Kwak, G. (2014). Room temperature fluorescent conjugated polymer gums, *Adv. Funct. Mater.*, **24**, pp. 1928–1937.
41. Jin, Y., Yoon, J., Sakaguchi, T., Lee, C., and Kwak, G. (2016). Highly emissive, water-repellent, soft materials: Hydrophobic wrapping and fluorescent plasticizing of conjugated polyelectrolyte via electrostatic self-assembly, *Adv. Funct. Mater.*, **26**, pp. 4501–4510.
42. Pan, C., Sugiyasu, K., Wakayama, Y., Sato, A., and Takeuchi, M. (2013). Thermoplastic fluorescent conjugated polymers: Benefits of preventing π–π stacking, *Angew. Chem. Int. Ed.*, **52**, pp. 10775–10779.
43. (a) Pisula, W., Kastler, M., Wasserfallen, D., Pakula, T., and Müllen, K. (2004). Exceptionally long-range self-assembly of hexa-peri-hexabenzocoronene with dove-tailed alkyl substituents, *J. Am. Chem. Soc.*, **126**, pp. 8074–8075; (b) Kastler, M., Pisula, W., Wasserfallen, D., Pakula, T., and Müllen, K. (2005). Influence of alkyl substituents on the solution- and surface-organization of hexa-peri-hexabenzocoronenes, *J. Am. Chem. Soc.*, **127**, pp. 4286–4296.
44. Giri, N., Del Pópolo, M. G., Melaugh, G., Greenaway, R. L., Rätzke, K., Koschine, T., Pison, L., Gomes, M. F. C., Cooper, A. I., and James, S. L. (2015). Liquids with permanent porosity, *Nature*, **527**, pp. 216–221.
45. (a) Ogoshi, T., Aoki, T., Shiga, R., Iizuka, R., Ueda, S., Demachi, K., Yamafuji, D., Kayama, H., and Yamagishi, T. (2012). Cyclic host liquids for facile and high-yield synthesis of [2]rotaxanes, *J. Am. Chem. Soc.*, **134**, pp. 20322–20325; (b) Ogoshi, T., Aoki, T., Ueda, S., Tamura, Y., and Yamagishi, T. (2014). Pillar[5]arene-based nonionic polyrotaxanes and a topological gel prepared from cyclic host liquids, *Chem. Commun.*, **50**, pp. 6607–6609.

Chapter 2

Rigid–Flexible Hybrid Design for Photofunctional Molecules and Materials

Shohei Saito

Graduate School of Science, Kyoto University, Kitashirakawa-Oiwake, Sakyo, Kyoto 606-8502, Japan
s_saito@kuchem.kyoto-u.ac.jp

Photophysical properties of organic materials are controlled by π-conjugated structures. Since π-structures generally consist of rigid aromatic rings and multiple bonds, their conformational variety is relatively limited. In this chapter, flapping molecules and their versatile functions are introduced, whose structure is composed of rigid aromatic wings and a conformationally flexible joint. Photophysical properties and molecular assembly modes can be tailored in a series of the flapping systems. Applications are also described particularly as a viscosity probe and a light-melt adhesive.

Light-Active Functional Organic Materials
Edited by Hiroko Yamada and Shiki Yagai
Copyright © 2019 Jenny Stanford Publishing Pte. Ltd.
ISBN 978-981-4800-15-0 (Hardcover), 978-0-429-44853-9 (eBook)
www.jennystanford.com

2.1 Introduction

2.1.1 Hybridization of Rigidity and Flexibility

Conformational rigidity plays an important role in the molecular chemistry and materials science of π-conjugated systems. However, the inherent rigidity of π-skeletons means the difficulty in providing convertible photophysical properties, which would originate from the flexibility of molecular structures. If conformational flexibility is effectively combined with rigid π-conjugated molecules composed of aromatic rings, new molecule-based technologies can be explored by using dynamic motions of the designed π-skeletons. In other words, construction of a hybrid π-system that combines rigidity and flexibility will provide new functional molecules and materials, for example, versatile systems featuring luminescent properties and π-stacking ability based on the *structural rigidity* and, at the same time, convertible photophysical properties derived from the *conformational flexibility*.

2.1.2 Controlling Photophysical Properties by Conformation Change

Although conformational flexibility would be the key to create dynamic functions, not every flexible π-molecule displays convertible photophysical properties. In general, conformational changes in π-molecules do not necessarily lead to significant perturbation of electronic properties. To construct characteristic systems whose properties change in conjunction with the conformational changes, sophisticated molecular design is required. Photochromic compounds such as azobenzene [1–4], diarylethene [5, 6], and spiropyran [7] have been studied as successful examples to switch photophysical properties and to induce mechanical motions in films and crystals. Their photoresponses derive from photoisomerization reactions involving recombination of chemical bonds. On the other hand, a well-studied example of the conformationally flexible functional system is twisted intramolecular charge transfer (TICT) molecules, which are constituted of electron donor and acceptor moieties connected with a rotatable single bond [8, 9]. Photoexcitation of TICT molecules induces an intramolecular electron transfer from the donor to the acceptor. In this chapter, these photochromic compounds and TICT systems are not highlighted, since other reviews can be referred [1–9].

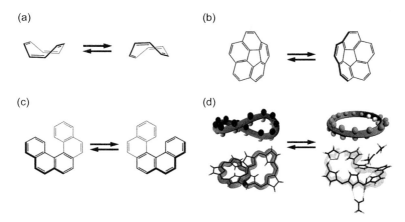

Figure 2.1 Conformational flexibility of π-conjugated molecules: (a) inversion of cyclooctatetraene (COT), (b) inversion of corannulene, (c) flipping of [5] helicene, and (d) Hückel–Möbius topological switch of an expanded porphyrin.

This chapter will focus on the conformational flexibility of parent π-molecules without bond breakage or bond formation, which can be found in the following representative examples (Fig. 2.1), namely, (a) inversion of tub-shaped cyclooctatetraene (COT) [10], (b) inversion of bowl-shaped π-systems such as corannulene [11] and sumanene [12], (c) flipping of helical π-systems such as helicenes [13] and twisted perylene bisimides [14], and (d) Hückel–Möbius topological switch of expanded porphyrins [15], which are flexible π-macrocycles. Practical application of these conformational flexibilities to produce useful functions is still challenging. In particular, COT and expanded porphyrins are characteristic, in which the electronic structure significantly depends on their molecular conformation because of the emergence of specific electronic states such as 4nπ Hückel antiaromaticity [16] and 4nπ Möbius aromaticity [17] in the ground state, and 4nπ Baird aromaticity in the excited state [18].

2.1.3 Flapping Fluorophore

On the basis of the background mentioned earlier, a new hybrid π-system holding rigidity and flexibility has been designed, where two rigid anthraceneimide fluorophores are fused with a flexible COT ring (Fig. 2.2). A set of the rigid–flexible hybrid π-systems is named FLAP (an abbreviation of flexible and aromatic photofunctional systems) [20–25]. **FLAP1** is the first fluorescent molecule that can emit an environment-dependent RGB (red, green, and blue) luminescence as a single component without changing the excitation wavelength [20]. **FLAP1** shows green fluorescence (FL) in solution, blue FL in a polymer film, and red FL in a crystalline phase (Fig. 2.3). Although some compounds have been reported to show multi-color luminescence [26–31], **FLAP1** has a unique mechanism of the multiple emission. Conventional mechanisms of luminescence color changes are typically based on an ON/OFF switching of the following photochemical processes by external stimuli or environmental changes [32–34]: intramolecular charge transfer, including TICT mechanism [8, 9]; excited-state intramolecular proton transfer (ESIPT) [35, 36]; fluorescence resonance energy transfer (FRET) [37, 38]; excimer/exciplex formation [39, 40]; J-aggregation [41]; aggregation-induced emission (AIE) by restriction of intramolecular rotation [42]; phosphorescence through intersystem crossing [43, 44]; and d-orbital interaction of metal complexes particularly involving Au···Au [45, 46] and Pt···Pt [47, 48] contacts in the excited state. Recently, thermally activated delayed fluorescence (TADF) [49, 50] and upconversion FL via triplet–triplet annihilation (TTA-UC) [51, 52] have also attracted attentions as unique FL mechanisms. In contrast, mechanism of the dual FL of **FLAP1** is different from these cases. The dual FL of **FLAP1** arises from a conformational change in a single π-system that is not accompanied by charge transfer. The excited-state dynamics is rather similar to those of dibenzo[b,f]oxepine [53] and N,N'-disubstituted dihydrodibenzo[a,c] phenazines [54, 55]. It is still under investigation whether excited-state aromaticity would contribute to their planarization dynamics in the lowest singlet excited state (S_1), but our recent study indicates that fused π-systems showing the excited-state planarization do not necessarily display excited-state aromaticity [25].

Introduction | 31

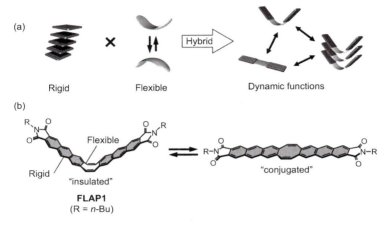

Figure 2.2 (a) Hybridization of rigidity and flexibility for designing FLAP molecules and (b) an example of FLAP molecule (**FLAP1**) composed of rigid anthraceneimide wings and a flexible cyclooctatetraene joint.

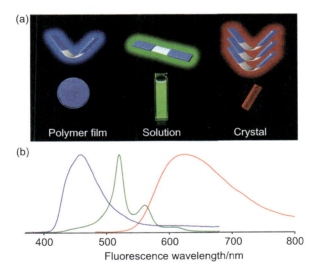

Figure 2.3 Single-component RGB luminescence of **FLAP1** depending on the surrounding environments. (a) Emissive molecular structures and photographs. (b) FL spectra of **FLAP1** in each environment, namely, blue FL in a doped polymer film (left), green FL in an organic solvent (middle), and red FL of single crystals (right).

FLAP1 is most stable in a bent conformation bearing a V-shaped geometry at the S_0 ground state, and the energy barrier of the conformational inversion is estimated to be lower than 10 kcal/mol. Therefore, it undergoes a fast flapping behavior at room temperature. In the flat form, $S_0 \rightarrow S_1$ excitation energy becomes significantly smaller than that in the bent form. Consequently, the flat form is energetically more stable in S_1 (Fig. 2.4). Photoexcitation of **FLAP1** changes its conformation in solution from a V-shaped form into a flat form in S_1, emitting green FL. In contrast, under highly viscous media such as in polymer matrix and in frozen glass, the V-shaped conformer cannot relax in S_1, and thus blue FL is emitted from almost retained V-shaped form. In other words, **FLAP1** shows viscosity-dependent dual FL in dispersed conditions (see Sec. 2.2) [23]. On the other hand, despite its nonplanar structure, **FLAP1** can construct a twofold columnar packing structure (Fig. 2.3a). In a crystalline phase, **FLAP1** shows an excimer-like red FL because of significant intermolecular interaction in S_1. Even in kinetically formed amorphous phase, a yellow to orange FL is observed in the solid state, which enables instant identification of supramolecular aggregations of **FLAP1** by the FL color change. Crystal packing structures as well as solid-state FL spectra can be controlled by changing the bulkiness of the terminal substituents in **FLAP1**. It has turned out that the remarkable red shift observed in the crystals of **FLAP1** bearing n-butyl substituents originates from the unique twofold columnar π-stacking structure of the V-shaped molecules [20].

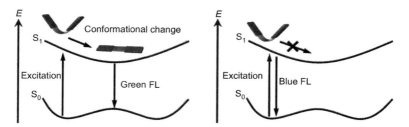

Figure 2.4 Schematic energy diagram of **FLAP1**. (a) Green FL observed in low-viscosity media and (b) blue FL observed in high-viscosity media.

2.2 Viscosity Mapping Technique

Viscosity mapping by synthetic fluorescent probes is a useful technique because it can visualize local viscosity distribution in heterogeneous media. Although the local viscosity analyzed by chemical probes does not simply correspond to bulk viscosity measured by viscometers and rheometers, these properties are closely related in homogeneous solvents, in particular Newtonian fluids. By virtue of spatial heterogeneity mapping, molecular viscosity probes have an advantage over these instruments [56–64].

2.2.1 Molecular Design of Fluorescent Viscosity Probes

Molecular rotors are most representative viscosity probes. In highly viscous media, an internal bond rotation in S_1 is suppressed, so that the FL quantum yield of fluorogenic molecular rotors increases (Fig. 2.5). These fluorophores have been used for the study of microenvironments in polymeric materials [63, 64], artificial membranes, and cell organelles [56–62].

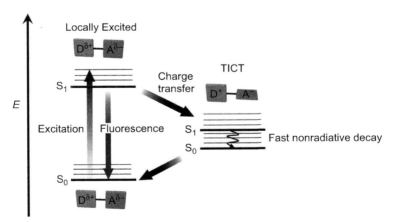

Figure 2.5 Excited-state dynamics of fluorogenic viscosity probe. Fluorescence quenching by fast nonradiative decay in a twisted conformation, which is suppressed in highly viscous media with enhancement of the fluorescence quantum yield.

Local viscosity has been quantitatively analyzed using the *ratiometric fluorescence technique* [65, 66], in which the distribution of local viscosity can be mapped based on an intensity ratio of two FL bands at different wavelengths. Recently, *fluorescence lifetime imaging* (FLIM) [67–71] has also been an important alternative to quantitative viscosity investigation, which enables visualization of cellular and materials heterogeneity based on viscosity-dependent FL lifetime. In addition to the advancement in imaging techniques, a variety of probe molecules showing viscosity-sensitive FL with small polarity dependence [72] have been synthesized and widely used, such as CVJ (cyanovinyljulolidine), BODIPY (boron-dipyrromethene), and Cy (cyanine) derivatives (Fig. 2.6) [56–62]. However, structural design of these viscosity probes still relies on molecular rotors featured by intramolecular rotational dynamics in S_1, including bond twists that result in *trans–cis* photoisomerization. Flapping fluorophore **FLAP2** and **FLAP4** in Fig. 2.7 show a polarity-independent ratiometric FL feature as a set of novel viscosity probes [23]. The structural design is based on the hybridization of a flexible COT and rigid fluorescent aromatic wings. It is noteworthy that some other viscosity probes with different structural designs have been recently reported, for example, phosphorescent molecular butterfly bearing a Pt–Pt contact [73] and amine-introduced benzoborole showing B–N bond-cleavage-induced intramolecular charge transfer (BICT) [74].

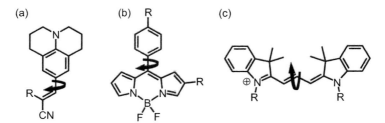

Figure 2.6 Molecular rotors based on (a) CVJ (cyanovinyljulolidine), (b) BODIPY (boron-dipyrromethene), and (c) Cy (cyanine) skeletons. These molecules work as viscosity probes by sensing local environment through an internal rotation indicated by arrows.

2.2.2 Flapping Viscosity Probe

2.2.2.1 Fluorescence and excited-state dynamics

FLAP2 has been synthesized by acene elongation of a COT-fused naphthalene dimer (Fig. 2.7a) [75]. **FLAP2** shows intense green FL at around 520, 560, and 610 nm in common organic solvents. FL quantum yield (Φ_f) is $\Phi_f = 0.34$ in THF. Large Stokes shift (4580 cm^{-1}) indicates a large conformational change in S_1. At room temperature, a weak broad FL band is also observed in the blue region (450–500 nm). In a frozen glass of 2-methyl tetrahydrofuran (MTHF) media at 77 K, the green emission bands disappear and the blue emission bands increase in intensity (Fig. 2.7). **FLAP3**, bearing an extra methyl group at the COT ring, shows a distinct FL behavior from **FLAP2**. Namely, **FLAP3** displays blue FL at around 440 and 465 nm ($\Phi_f = 0.28$ in THF). The Stokes shift of **FLAP3** (1200 cm^{-1}) is much smaller than that of **FLAP2**, indicating that the V-shaped to planar conformational change in S_1 is significantly suppressed by the small steric hindrance on the COT ring; that is, the excited-state dynamics of the FLAP systems is featured by the conformational flexibility of COT. Calculated energy profiles explain the distinct FL properties of **FLAP2** and **FLAP3**. In the most stable V-shaped conformation in S_0, the bond alternation around the *cis*-olefins of the COT ring is explicit, meaning that the two anthraceneimide moieties are not effectively π-conjugated. The S_1 energy diagram of **FLAP2** indicates two different energy minima at a shallow V-shaped conformation and a flat one. The energy barrier between the two minima is small enough to get over at room temperature, and the flat conformation is slightly more stable in S_1 (Fig. 2.8b).

As a result, the intense green FL at 520 nm of **FLAP2** can be assigned to the emission from the flat conformation, while the weak blue FL observed in 450–500 nm can be assigned to the emission from the shallow V-shaped conformation. According to molecular orbitals analysis, less effective π-conjugation through the COT joint is confirmed in the shallow V-shaped conformation, while an electronic contribution of COT becomes significant in the flat conformation. In sharp contrast, the S_1 energy diagram of **FLAP3** has a single minimum at a V-shaped conformation, supporting a steric hindrance of the methyl group on COT for the conformational planarization in

S_1. The blue FL of **FLAP3** with a smaller Stokes shift is assigned to the emission from the V-shaped structure. Note that the FL behavior of **FLAP2** is similar to that of dibenzo[b,f]oxepin [76], whose excited-state 8π-aromaticity has been predicted in S_1 [19].

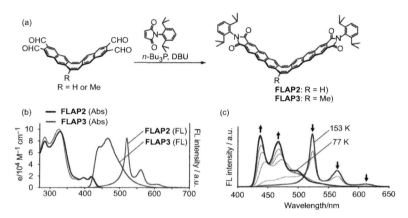

Figure 2.7 (a) Absorption and fluorescence spectra of **FLAP2** and **FLAP3**, and (b) the spectral change in **FLAP2** in fluorescence at lower temperature.

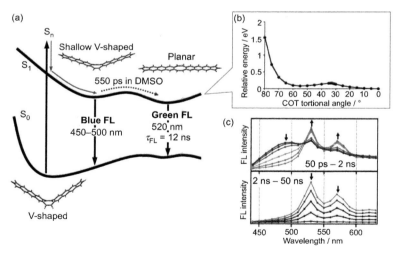

Figure 2.8 Energy diagram of **FLAP2** in S_0 and S_1. (a) Relaxed scan profile of **FLAP2** in S_1 at the cam-B3LYP/6-31+G* level, in which the terminal substituents are replaced by hydrogen atoms, and (b) time-resolved fluorescence spectroscopy of **FLAP2**.

Conformational dynamics of **FLAP2** in S_1 has been supported by time-resolved FL and IR spectroscopies [23]. The broad blue FL from V-shaped form is observed immediately after the pulse excitation (Fig. 2.8c). This blue FL band decreases with a time constant of 550 ps in DMSO, and the intense green FL bands from the flat form concomitantly increase around 530 and 570 nm, whose decay constant is ~12 ns. Time-resolved IR spectroscopy of **FLAP2** has supported this result. As a result, the dynamics of **FLAP2** in S_1 has been elucidated, as shown in Fig. 2.8a.

2.2.2.2 Polarity-independent viscochromism

A quantitative study of the viscochromism in the FLAP system has been demonstrated (Fig. 2.9). A mixed solvent of DMSO/glycerol affords suitable Newtonian fluid for this purpose, namely, media viscosity is independent of shear rate [77]. By changing the ratio of DMSO/glycerol, the viscosity can be changed without altering polarity. Hydrophilic **FLAP4** shows good solubility in these media. Importantly, FL spectra of **FLAP4** as well as **FLAP2** is independent of the polarity. The green FL band displays a negligible shift in various organic solvents such as toluene, tetrahydrofuran (THF), dichloromethane (DCM), dimethyl formamide (DMF), acetonitrile (MeCN), and dimethyl sulfoxide (DMSO). These solvents cover a wide range of relative dielectric constants ε_r from 2.4 to 46.7 at room temperature, while the viscosity is confined within a small range of 0.4–2.2 cP. The FL spectral shape of **FLAP4** is almost preserved despite the wide range of relative dielectric constants, meaning that the emissive planar structure in S_1 is not charge polarized as supported by calculations.

On the other hand, a large viscosity dependence of **FLAP4** in FL has been demonstrated. As the viscosity increases, the relative intensity of the blue FL band (450–500 nm) becomes higher in comparison with the green FL band (525 nm). On the basis of the Förster–Hoffmann rule [57, 78], the FL intensity ratio (I_{461}/I_{525}) of the blue FL band (461 nm) and the green FL band (525 nm) has been plotted with the viscosity in a double logarithmic plot, and a linear relationship in a viscosity range of 2.2 to 100 cP has been confirmed as a calibration line. FLAP has a higher sensitivity in this viscosity range than the conventional viscosity probes based on molecular rotors. Namely, the FL spectral shape is more largely dependent on the viscosity, presumably because the flapping motion of **FLAP4** requires a larger volume to displace the surrounding solvent molecules as compared with the rotational motion of typical molecular rotors [79]. The polarity-independent ratiometric viscochromism of FLAP enables real-time visualization of a local viscosity change without sensing the local polarity environment.

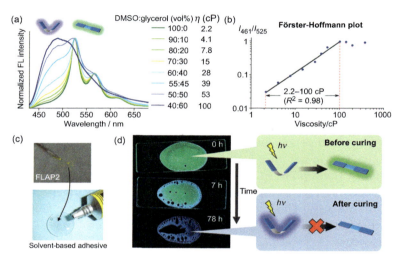

Figure 2.9 Viscochromism of **FLAP2**. (a) Viscosity-dependent FL spectra of **FLAP2**. (b) Double logarithmic plot of FL intensity ratio I_{461}/I_{525} versus viscosity measured by a viscometer. (c) Doping **FLAP2** into a solvent-based adhesive. (d) Real-time monitoring of the curing process of a solvent-based adhesive.

2.2.2.3 Monitoring the epoxy resin curing

By using the flapping viscosity probe **FLAP2**, epoxy resin curing can be monitored in a real-time manner [23]. In this curing process, both the local viscosity and polarity must increase because the curing reaction generates polar hydroxy groups and because the microviscosity increases with decrease in the free volume. To prove the efficacy of FLAP, an epoxy prepolymer of bisphenol A diglycidyl ether (BADGE), pentaerythritol tetrakis(2-mercaptoacetate) (PETM), and n-Bu$_3$N have been used as an epoxy agent, a hardener, and a catalytic accelerator, respectively (Fig. 2.10). These reagents are cured in two steps: First, the BADGE epoxy prepolymer is mixed

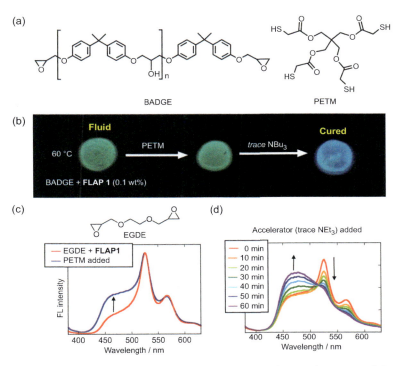

Figure 2.10 (a) Chemical structure of an epoxy prepolymer of BADGE and the hardner of PETM. (b) FL color change of curing epoxy resin (BADGE + PETM) doped with **FLAP1**. (c) FL spectral change of curing epoxy resin (EGDE + PETM) doped with **FLAP1**.

with the PETM hardener, and second, a catalytic amount of n-Bu$_3$N accelerates the crosslinking polymerization between BADGE and PETM. Before curing, a detectable amount of **FLAP2** (0.1 wt.% ratio) is doped into the BADGE epoxy prepolymer, which shows green FL at 60°C. After the addition of the PETM hardener to the BADGE epoxy prepolymer, the green FL is still maintained, but the FL color largely changes into blue after the addition of n-Bu$_3$N, which means that the conformational flattening of **FLAP2** is eventually suppressed in S$_1$. With another epoxy agent, ethylene glycol diglycidyl ether (EGDE), spectroscopic monitoring at room temperature has been realized, and it has turned out that the curing process takes 1 h (Fig. 2.10). Other dual fluorescent molecules with the mechanisms of TICT [8, 9] and ESIPT [35, 36] cannot be used for this purpose. A TICT-type dye 4-(dimethylamino)benzonitrile does not work as a viscosity probe, and it works rather as a polarity probe because the charge-polarization is significant in S_1. In an ESIPT-type dye, 2-(2′-hydroxyphenyl)oxazole, the structural change during the ESIPT process would be too small to be suppressed by adhesive curing, and the FL band with a large Stokes shift remains even after the complete epoxy curing takes place. As a result, the competence of FLAP has been demonstrated as a polarity-independent ratiometric fluorescent probe.

2.3 Light-Removable Adhesive

Photoinduced phase transition has been well studied on materials containing isomerization dyes such as azobenzene [1–4], diarylethene [5, 6], and spiropyran [7]. The study has attracted much attention in materials science, because optical properties and mechanical motion of the materials can be controlled. In industry, photocurable resins are widely used for adhesion, coating, and sealing. A unique example is a dicing tape [80], which is commercially used to attach semiconductor wafers temporarily during a dicing process in order not to be blown off. The working mechanism of a dicing tape is based on the induction of detachment by a photocuring process. However, there are limited reports on "rigid materials that dissolve [81–85] or soften [86–88] by light irradiation," which would be new light-removable adhesives with many potential applications. In the following sections, recently reported functional materials contributing to this field are introduced, and the light-melt adhesive based on FLAP is described in detail.

2.3.1 Polymer and Supramolecular Approach

Some polymer materials are developed as photodeactivatable resins based on several operation principles, such as photoinduced crosslinking [89, 90], photoacid-catalyzed chain modification [91], photocleavage of polymer chains [92, 93], and photothermal cleavage of supramolecular chains [94]. For example, photodimerization of coumarine units can be utilized for crosslinking of polymer chains, inducing deactivation of a pressure-sensitive adhesive (Fig. 2.11a) [90]. Diarylethene-introduced polymer material shows Diels–Alder activity when the diarylethene unit takes the open form (Fig. 2.11b) [93]. Dissociation of the polymer chains is induced by visible light irradiation when the diarylethene-containing polymer taking the closed form is warmed at high temperature, which results in a decrease in adhesive strength. Adhesion switching has been also reported with a host–guest complex of cyclodextrin (CD) and azobenzene (Azo) at interfaces of polymer gels (Fig. 2.11c) [92]. When Azo takes the *trans* form, host–guest interaction works for bonding two gels. Upon UV irradiation, Azo changes into the *cis* form, which is released from the CD host because of its bent structure. As a result, the host–guest interactions at the interface disappear, and therefore two gels are spontaneously separated.

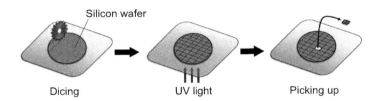

Figure 2.11 Dicing tape for temporary attachment of a silicon wafer. After dicing the wafer, detachment of the silicon pieces is induced by UV irradiation.

2.3.2 Liquid Crystal Approach

Photoresponsive LC is a suitable platform for realizing a quick photoinduced isothermal phase transformation into a fluid mixture [95], as early examples represented in Fig. 2.13. Until recently, however, an application of LC materials for removable adhesives has not been explored [83, 96]. The difficulty comes from the requirement for satisfying both sufficient bonding strength and its rapid disappearance by photoirradiation.

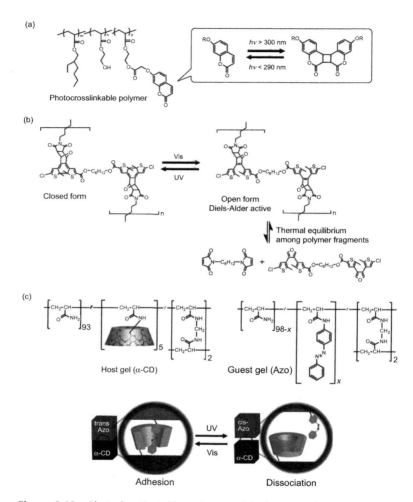

Figure 2.12 Photodeactivatable polymers. (a) Photocrosslinkable pressure-sensitive adhesive (PSA), (b) diarylethene-based Diels–Alder active polymer, and (c) host–guest complex gel constituted of cyclodextrin (CD) and azobenzene (Azo) units.

2.3.3 Light-Melt Adhesive

Photomelting columnar LC bearing a FLAP mesogen is expected to satisfy the requirements for the practical application of light-melt adhesive. V-shaped FLAP molecules are stacked in a columnar manner to display sufficient cohesive strength even under high temperature, but its bonding ability is immediately lost by UV irradiation triggering photodimerization of the anthracene moiety at the FLAP wings [22].

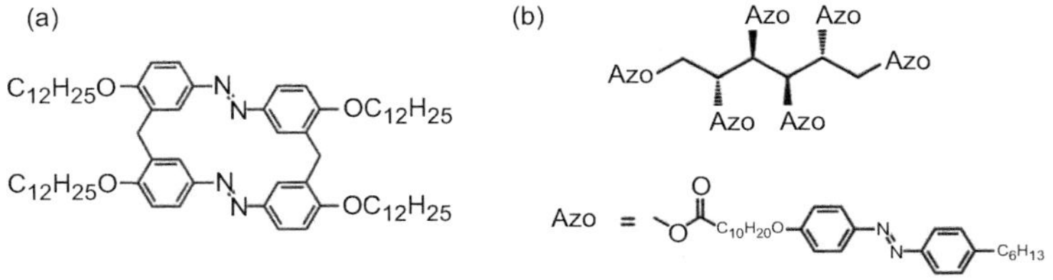

Figure 2.13 Early examples of molecule-based photomelting materials. (a) A cyclic azobenzene dimer and (b) an azobenzene oligomer.

2.3.3.1 Requirements for applications

Hot-melt adhesives are widely used in industry as a temporary adhesive. These adhesives are composed of materials that melt at high temperature. Therefore, high-temperature-resistant bonding is a challenging issue for temporary adhesives. In this context, the light-melting function offers a novel manufacturing technique if the following requirements are fully satisfied. First, adequate strength for a temporary bond is required even under heating conditions. Standard temporary bonding needs 1 MPa adhesive strength, while permanent adhesion in some applications demands more than 10 MPa. Second, a significant reduction in the bonding strengths must be induced by light irradiation. Third, rapid photoresponse, ideally within a few seconds, is necessary for efficient separation technology in a manufacturing process. The required wavelength of the excitation light depends on each application, but instruments for 365 nm UV irradiation are widely equipped in laboratories and factories where they use UV-curable resins [97].

The aforementioned requisites have been fulfilled by the light-melt adhesive described below: (1) a shear strength over 1 MPa up to 110°C for bonding glass plates, (2) an 85% decrease in the shear strength by UV irradiation, and (3) instant photomelting of the LC film in a few seconds has been realized (Fig. 2.14). In addition, the phase transition is reversible, and so the light-melt adhesive can be used as a reworkable adhesive. The photoinduced isothermal phase transformation between the LC and melted phases is accompanied by a color change in FL, which visualizes a difference in the bonding/nonbonding phase in a contactless fashion [22].

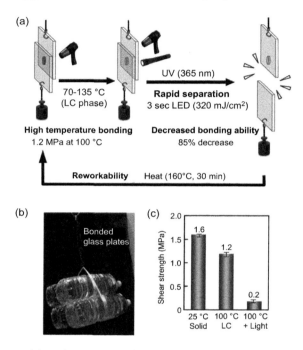

Figure 2.14 (a) Performance of the light-melt adhesive, a liquid crystal of **FLAP5** in Fig. 2.15. (b) Demonstration of strong bonding properties. A few milligram of the adhesive is sandwiched between two glass plates, hanging 8 kg water bottles. (c) Ultimate shear strength of the sample specimen in each condition.

2.3.3.2 Materials design

Light-melt adhesive is made of the LC material **FLAP5**. Molecular design is based on the FLAP mesogen bearing typical dendritic carbon chains (Fig. 2.15a). The hybrid design of rigid anthracene and flexible COT affords two important characteristics to this material. First, the rigid anthracene arms of the V-shaped FLAP display strong stacking ability to construct a columnar array in the condensed phases, which results in high cohesive force of the LC material and thus high-temperature-resistant bonding. Second, the flexible COT ring changes a conformation of FLAP into a flat one by photoexcitation in the LC phase, leading to the photodimerization of the anthracene moiety. This photoreaction induces the photomelting function, accompanied by detachment of bonded glass plates.

FLAP5 has a rectangular columnar LC phase between 65 and 140°C, in which the V-shaped molecules align on top of each other and the π-stacked arrays are located side by side. Construction of the columnar π-stacking is supported in a single-crystal X-ray structure of the corresponding analogue **FLAP6**, which has no peripheral chains (Fig. 2.15b). Interfacial distance $d_{\pi-\pi}$ between the stacked anthracene arms is 3.50 Å, indicating the formation of standard slipped π-stacking at both arms. The strong intermolecular interaction and high cohesive force of **FLAP5** are also confirmed by a large enthalpy change in a phase transition between the columnar LC and isotropic liquid. The enthalpy change over 30 kJ/mol is one of the largest values among LCs that do not involve hydrogen bond interaction [98].

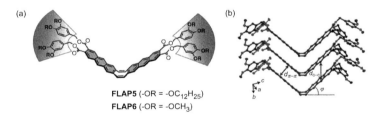

FLAP5 (-OR = -OC$_{12}$H$_{25}$)
FLAP6 (-OR = -OCH$_3$)

Figure 2.15 Molecular structures of **FLAP5** and its analogue **FLAP6** (left). Twofold π-stacking structure of V-shaped **FLAP6** molecules (right).

2.3.3.3 Adhesive performance

A thin film of **FLAP5**, sandwiched with two glass plates, gets melt by 365 nm UV-LED irradiation. Observation by polarized optical microscopy (POM) and thermography indicates a photoinduced isothermal phase transformation into an isotropic liquid. The nature of the LC phase is important to induce the photomelting function, since this transformation has not been observed in the solid phase. The resulting isotropic fluid contains a photodimer and trace of oligomers of **FLAP5** produced by UV irradiation, while unreacted monomer also remains in the mixture. In a 130-μm-thick adhesive thin film, half of the monomers are dimerized until glass plates are separated by light. Complete consumption of monomers is not necessary for the photomelting event probably because the destruction of the columnar LC structure is spontaneously induced by amplification of a disorganized phase [1]. Less symmetric molecular structure of the isolated photodimer implies a photomelting mechanism that in situ photodimer generation destabilizes the LC phase of **FLAP5** due to the unsuitable molecular shape for the columnar packing (Fig. 2.16).

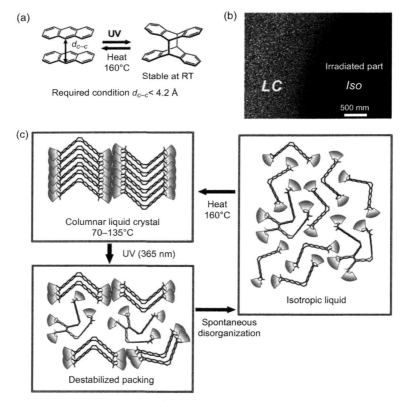

Figure 2.16 (a) Photodimerization of anthracene. (b) POM image of photomelting LC constituted of **FLAP5**. (c) Photomelting mechanism of **FLAP5**.

The light-melt adhesive has a practical shear strength (more than 1 MPa) for temporary bonding. While the adhesive film displays high-temperature-resistant bonding properties, the glass plates are separated within a few seconds by UV irradiation. Ultimate shear strengths are 1.6 MPa at 25°C, 1.2 MPa at 100°C, 1.1 MPa at 110°C, and 0.9 MPa at 120°C, in which a 130-μm-thick film of **FLAP5** is sandwiched by non-treated glass plates. Generally speaking, adhesive strength is discussed in relation with cohesion force and adhesion force. The cohesion force is an internal strength of the adhesive material, which can be intimately associated to an intermolecular interaction particularly in the case of small molecules. On the other hand, the adhesion force is an interaction between adhesive materials and a substrate surface. If $F_{cohesion} > F_{adhesion}$, the bonding strength is determined by the adhesion force, and thus it largely depends on the conditions of substrate surfaces such as hydrophilicity. When $F_{cohesion} < F_{adhesion}$, in contrast, the bonding strength is determined by the cohesion force regardless of the surface conditions. Shear strengths of the sample specimen using the light-melt adhesive (**FLAP5**) are confirmed to be independent of hydrophilicity of glass surface. Therefore, the cohesion force plays a key role in determining the shear strength of the test specimen rather than the adhesion force (Fig. 2.17) [97].

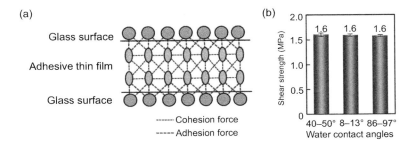

Figure 2.17 (a) Cohesion force and adhesion force in an adhesive specimen. (b) Constant shear strength regardless of hydrophilicity of the glass surface, measured for the bonded specimens using **FLAP5**.

The shear strength remarkably reduces to 0.2 MPa after 320 mJ/cm² UV exposure at 100°C. Quick separation has been achieved in a few seconds by using 365 nm UV-LED with

160 mW/cm irradiance. Since the condensed material of **FLAP5** efficiently absorbs UV light, the photoinduced separation takes place around an interface between the adhesive film and the irradiated glass plate. The light transmittance exponentially decreases according to the thickness of the **FLAP5** films. More than 95% of the 365 nm light is absorbed within 3 μm depth from the interface. The total irradiation dose required for the glass separation is almost constant regardless of the film thickness over 5 μm (Fig. 2.18). Therefore, only a small amount of adhesive residue remains on the irradiated glass plate. In addition, the residue can be easily removed by common organic solvents due to the good solubility of **FLAP5**.

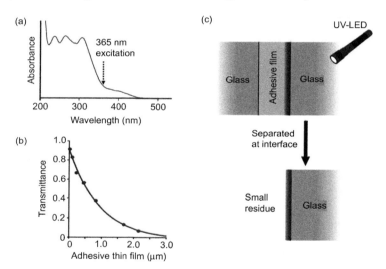

Figure 2.18 (a) Absorption spectrum of the LC film composed of **FLAP5**. (b) Transmittance of the 365 nm excitation light depending on the film thickness. (c) Photoinduced detachment near the interface between the adhesive and the glass plate.

Thin films of **FLAP5** work as a reworkable adhesive. When the photomelted mixture is heated at 160°C for 15 min, both the shear strength around 1.5 MPa at 25°C and the rapid photomelting function at 100°C can be recovered. Furthermore, an FL color reflects the phase of materials, namely, bonding or nonbonding state. Blue FL is observed when the adhesive film gets melted, while green FL in the

LC phase is a sign of the adhesive recovery by thermal back reaction of the photodimer (Fig. 2.19).

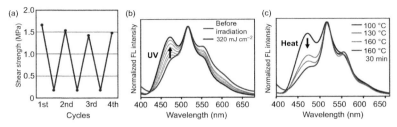

Figure 2.19 (a) Repeated cycles of photomelting and thermal recovery. FL spectral change in an LC film of **FLAP5** during (b) UV irradiation and (c) thermal back reaction.

2.3.3.4 Working mechanism

The photomelting mechanism has been demonstrated in consideration of the dynamics of **FLAP5** in the singlet excited state (S_1) (Fig. 2.20). Photodimerization of anthracene moieties takes place in S_1 [99] if the reactive sites can approach each other in the condensed phase. In the literature, the required distance d_{C-C} is reported to be within 4.2 Å in order to form new C–C bonds between the central carbon atoms of the anthracene units [100]. In spite of the tight π-stacking of the anthracene arms with $d_{\pi-\pi} \sim 3.5$ Å, the corresponding distance d_{C-C} in the columnar packing of **FLAP5** is estimated to be ca. 4.7 Å, which originates from the slipped alignment of the anthracene arms. As a result of the longer d_{C-C} distance, photodimerization is not allowed at room temperature, in which the conformation of **FLAP5** is fixed in the solid phase. In contrast, the columnar LC structure can be perturbed by the photoexcitation of **FLAP5** because the conformational planarization in S_1 described above is allowed in the LC phase, and the dimerization occurs when the d_{C-C} distance gets closer. The dynamic conformational change in **FLAP5** in the LC phase has been elucidated by the fluorescence spectroscopy of the LC film. The LC film of **FLAP5** at 100°C shows green fluorescence with a large Stokes shift, which is the same as observed in its solution phase. The overall mechanism of the photomelting process has been interpreted as follows: Near the interface between the LC film of **FLAP5** and the irradiated glass substrate, the V-shaped molecules are photoexcited to induce the conformational planarization. As a result, the intermolecular distance d_{C-C} between the reactive carbon sites of the anthracene unit is significantly fluctuated. Photochemical dimerization takes place if an excited molecule in S_1 is coupled with a neighboring molecule in S_0, while other excited molecules relax back to S_0 accompanied by the green FL emitted from the planar conformation. The photoproducts play the role of "impurities" to induce disorganization of the columnar LC phase, which results in the photomelting event with a remarkable decrease in the bonding strength.

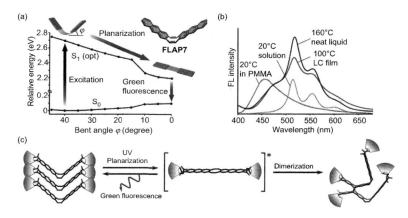

Figure 2.20 Photoresponse mechanism of **FLAP5** in the columnar LC phase. (a) Calculated energy profile of the parent structure **FLAP7**. (b) Fluorescence spectra of **FLAP5** in each phase of materials. (c) Conformational planarization in S_1 and subsequent dimerization of **FLAP5**.

In summary, a light-melt adhesive based on a photoresponsive columnar LC material has been developed. Tight π-stacking of the V-shaped FLAP molecules gives rise to sufficient bonding properties of the adhesive film even at high temperature. A remarkable photoinduced decrease in shear strength has been realized by the in situ dimerization of the LC molecule and the following disorganization of the columnar structure. Fast melting response achieves a quick separation of bonded glass plates leaving small adhesive residue. It has been envisioned that composite materials with the light-melt function will further improve performance in manufacturing processes, which will accelerate on-demand photoseparation technology complementary to the other switchable adhesion approaches [101].

2.4 Conclusion

Most applications using π-conjugated systems rely on the conformational rigidity of those molecular skeleton. It is still challenging to make use of conformational flexibility of molecules in materials science. In this chapter, molecular design, basic properties, and applications of conformationally flexible flapping molecules (FLAP) have been introduced. Flapping fluorophores enable polarity-independent ratiometric viscosity sensing, while a photoresponsive columnar LC featuring the flapping structure shows high performance as a light-melt adhesive. A series of FLAP based on a hybrid molecular design of rigidity and flexibility will develop new functional materials that show convertible photophysical and electronic properties in the condensed phase.

References

1. Ikeda, T. (2003) *J. Mater. Chem.*, **13**, pp. 2037–2057.
2. Browne, W. R. and Feringa, B. L. (2006) *Nature Nanotech.*, **1**, pp. 25–35.
3. Zhao, Y. and Ikeda, T. (Eds.) *Smart Light-Responsive Materials* (John Wiley & Sons, 2009).
4. White, T. J. and Broer, D. J. (2015) *Nature Mater.*, **14**, pp. 1087–1098.
5. Irie, M., Fukaminato, T., Matsuda, K., and Kobatake, S. (2014) *Chem. Rev.*, **114**, pp. 12174–12277.
6. Kobatake, S., Takami, S., Muto, H., Ishikawa, T., and Irie, M. (2007) *Nature*, **446**, pp. 778–781.
7. Klajn, R. (2014) *Chem. Soc. Rev.*, **43**, pp. 148.
8. Grabowski, Z. R., Rotkiewicz, K., and Rettig, W. (2003) *Chem. Rev.*, **103**, pp. 3899.
9. Sasaki, S., Drummen, G. P. C., and Konishi, G. (2016) *J. Mater. Chem. C*, **4**, pp. 2731–2743.
10. Paquette, L. A. (1993) *Acc. Chem. Res.*, **26**, pp. 57–62.
11. Wu, Y.-T. and Siegel, J. S. (2006) *Chem. Rev.*, **106**, pp. 4843.
12. Sakurai, H., Daiko, T., and Hirao, T. (2003) *Science*, **301**, 1878.

13. Shen, Y. and Chen, C.-F. (2012) *Chem. Rev.*, **112**, pp. 1463–1535.
14. Würthner, F. (2006) *Pure Appl. Chem.*, **78**, pp. 2341.
15. Saito, S. and Osuka, A. (2011) *Angew. Chem. Int. Ed.*, **50**, pp. 4342–4373.
16. Breslow, R. (1973) *Acc. Chem. Res.*, **6**, pp. 393.
17. Herges, R. (2006) *Chem. Rev.*, **106**, pp. 4820–4842.
18. Osuka, A. and Saito, S. (2011) *Chem. Commun.*, **47**, pp. 4330–4339.
19. Rosenberg, M., Dahlstrand, C., Kilså, K., and Ottosson, H. (2014) *Chem. Rev.*, **114**, pp. 5379–5425.
20. Yuan, C., Saito, S., Camacho, C., Irle, S., Hisaki, I., and Yamaguchi, S. (2013) *J. Am. Chem. Soc.*, **135**, pp. 8842.
21. Yuan, C., Saito, S., Camacho, C., Kowalczyk, T., Irle, S., and Yamaguchi, S. (2014) *Chem. Eur. J.*, **20**, pp. 2193–2200.
22. Saito, S., Nobusue, S., Hara, M., Tsuzaka, E., Yuan, C., Mori, C., Hara, M., Seki, T., Camacho, C., Irle, S., and Yamaguchi, S. (2016) *Nat. Commun.*, **7**, pp. 12094.
23. Kotani, R., Sotome, H., Okajima, H., Yokoyama, S., Nakaike, Y., Kashiwagi, A., Mori, C., Nakada, Y., Yamaguchi, S., Osuka, A., Sakamoto, A., Miyasaka, H., and Saito, S. (2017) *J. Mater. Chem. C*, **5**, pp. 5248.
24. Hada, M., Saito, S., Tanaka, S., Sato, R., Yoshimura, M., Mouri, K., Matsuo, K., Yamaguchi, S., Hara, M., Hayashi, Y., Röhricht, F., Herges, R., Shigeta, Y., Onda, K., and Miller, R. J. D. (2017) *J. Am. Chem. Soc.*, **139**, pp. 15792–15800.
25. Yamakado, T., Takahashi, S., Watanabe, K., Matsumoto, Y., Osuka, A., and Saito, S. (2018), *Angew. Chem. Int. Ed.*, **57**, 5438–5443.
26. He, G., Guo, D., He, C., Zhang, X., Zhao, X., and Duan, C. (2009) *Angew. Chem. Int. Ed.*, **48**, pp. 6132.
27. Yang, Y., Lowry, M., Schowalter, C. M., Fakayode, S. O., Escobedo, J. O., Xu, X., Zhang, H., Jensen, T. J., Fronczek, F. R., Warner, I. M., and Strongin, R. M. (2006) *J. Am. Chem. Soc.*, **128**, pp. 14081.
28. Zhao, Y., Gao, H., Fan, Y., Zhou, T., Su, Z., Liu, Y., and Wang, Y. (2009) *Adv. Mater.*, **21**, pp. 3165.
29. Ruff, Y., Buhler, E., Candau, S.-J., Kesselman, E., Talmon, Y., and Lehn, J.-M. (2010) *J. Am. Chem. Soc.*, **132**, pp. 2573–2584.
30. Sagara, Y. and Kato, T. (2011) *Angew. Chem. Int. Ed.*, **50**, pp. 9128.
31. Dong, Y., Xu, B., Zhang, J., Tan, X., Wang, L., Chen, J., Lv, H., Wen, S., Li, B., Ye, L., Zou, B., and Tian, W. (2012) *Angew. Chem. Int. Ed.*, **51**, pp. 10782.

32. Demchenko, A. P. (Ed.) (2010) *Advanced Fluorescence Reporters in Chemistry and Biology I, Fundamentals and Molecular Design*, Springer Series on Fluorescence 08 (Springer-Verlag, Berlin, Heidelberg).
33. Callan, J. F., Silva, A. P., and Magri, D. C. (2005) *Tetrahedron*, **61**, pp. 8551–8588.
34. Wu, J., Liu, W., Ge, J., Zhang, H., and Wang, P. (2011) *Chem. Soc. Rev.*, **40**, pp. 3483–3495.
35. Kwon, J. E. and Park, S. Y. (2011) *Adv. Mater.*, **23**, pp. 3615–3642.
36. Padalkar, V. S. and Seki, S. (2016) *Chem. Soc. Rev.*, **45**, pp. 169–202.
37. Fan, J., Hu, M., Zhan, P., and Peng, X. (2013) *Chem. Soc. Rev.*, **42**, pp. 29–43.
38. Yuan, L., Lin, W., Zheng, K., and Zhu, S. (2013) *Acc. Chem. Res.*, **46**, pp. 1462–1473.
39. Saigusa, H. and Lim, E. C. (1996) *Acc. Chem. Res.*, **29**, pp. 171–178.
40. Jenekhe, S. A. and Osaheni, J. A. (1994) *Science*, **265**, pp. 765–768.
41. Würthner, F., Kaiser, T. E., and Saha-Möller, C. R. (2011) *Angew. Chem. Int. Ed.*, **50**, pp. 3376–3410.
42. Mei, J., Leung, N. L. C., Kwok, R. T. K., Lam, J. W. Y., and Tang, B. Z. (2015) *Chem. Rev.*, **115**, pp. 11718–11940.
43. Zhao, Q., Li, F., Huang, C., (2010) *Chem. Soc. Rev.*, **39**, pp. 3007–3030.
44. Han, M., Tian, Y., Yuan, Z., Zhu, L., and Ma, B. (2014) *Angew. Chem. Int. Ed.*, **53**, pp. 10908–10912.
45. He, X. and Yam, V. W.-W. (2011) *Coord. Chem. Rev.*, **255**, pp. 2111–2123.
46. Seki, T. and Ito, H. (2016) *Chem. Eur. J.*, **22**, pp. 4322–4329.
47. Kato, M. (2007) *Bull. Chem. Soc. Jpn.*, **80**, pp. 287–294.
48. Ma, B., Li, J., Djurovich, P. I., Yousufuddin, M., Bau, R., and Thompson, M. E. (2005) *J. Am. Chem. Soc.*, **127**, pp. 28–29.
49. Uoyama, H., Goushi, K., Shizu, K., Nomura, H., and Adachi, C. (2012) *Nature*, **492**, pp. 235–240.
50. Tao, Y., Yuan, K., Chen, T., Xu, P., Li, H., Chen, R., Zheng, C., Zhang, L., and Huang, W. (2014) *Adv. Mater.*, **26**, pp. 7931–7958.
51. Singh-Rachford, T. N. and Castellano, F. N. (2010) *Coord. Chem. Rev.*, **254**, pp. 2560–2573.
52. Ye, C., Zhou, L., Wang, X., and Liang, Z. (2016) *Phys. Chem. Chem. Phys.*, **18**, pp. 10818.
53. Shukla, D. and Wan, P. (1993) *J. Am. Chem. Soc.*, **115**, pp. 2990–2991.

54. Zhang, Z., Wu, Y.-S., Tang, K.-C., Chen, C.-L., Ho, J.-W., Su, J., Tian, H., and Chou, P.-T. (2015) *J. Am. Chem. Soc.*, **137**, pp. 8509–8520.
55. Chen, W., Chen, C.-L., Zhang, Z., Chen, Y.-A., Chao, W.-C., Su, J., Tian, H., and Chou, P.-T. (2017) *J. Am. Chem. Soc.*, **139**, pp. 1636–1644.
56. Haidekker, M. A. and Theodrakis, E. A. (2016) *J. Mater. Chem. C*, **4**, pp. 2707–2718.
57. Kuimova, M. K. (2012) *Phys. Chem. Chem. Phys.*, **14**, pp. 12671–12686.
58. Haidekker, M. A. and Theodorakis, E. A. (2007) *Org. Biomol. Chem.*, **5**, pp. 1669–1678.
59. Klymchenko, A. S. (2017) *Acc. Chem. Res.*, **50**, pp. 366–375.
60. Klymchenko, A. S. and Kreder, R. (2014) *Chem. Biol.*, **21**, pp. 97–113.
61. Yang, Z., Cao, J., He, Y., Yang, J. H., Kim, T., Peng, X., and Kim, J. S. (2014) *Chem. Soc. Rev.*, **43**, pp. 4563–4601.
62. Kowada, T., Maeda, H., and Kikuchi, K. (2015) *Chem. Soc. Rev.*, **44**, pp. 4953–4972.
63. Itagaki, H., Horie, K., and Mita, I. (1990) *Prog. Polym. Sci.*, **15**, pp. 361–424.
64. Strehmel, B., Strehmel, V., and Younes, M. (1999) *J. Polym. Sci. Part B, Polym. Phys.*, **37**, pp. 1367–1386.
65. Haidekker, M. A., Brady, T. P., Lichlyter, D., and Theodorakis, E. A. (2006) *J. Am. Chem. Soc.*, **128**, pp. 398–399.
66. Fischer, D., Theodorakis, E. A., and Haidekker, M. A. (2007) *Nat. Protoc.*, **2**, pp. 227–236.
67. Suhling, K., Hirvonen, L. M., Levitt, J. A., Chung, P.-H., Tregidgo, C., Marois, A. L., Rusakov, D. A., Zheng, K., Ameer-Beg, S., Poland, A., Coelho, S., Henderson, R., and Krstajic, N. (2015) *Med. Photo.*, **27**, pp. 3–40.
68. Kuimova, M. K., Yahioglu, G., Levitt, J. A., Suhling, K. (2008) *J. Am. Chem. Soc.*, **130**, pp. 6672–6673.
69. Kuimova, M. K., Botchway, S. W., Parker, A. W., Balaz, M., Collins, H. A., Anderson, H. L., Suhling, K., and Ogilby, P. R. (2009) *Nature Chem.*, **1**, pp. 69–73.
70. Peng, X., Yang, Z., Wang, J., Fan, J., He, Y., Song, F., Wang, B., Sun, S., Qu, J., Qi, J., and Yan, M. (2011) *J. Am. Chem. Soc.*, **133**, pp. 6626–6635.
71. Yang, Z., He, Y., Lee, J.-H., Park, N., Suh, M., Chae, W.-S., Cao, J., Peng, X., Jung, H., Kang, C., and Kim, J. S. (2013) *J. Am. Chem. Soc.*, **135**, pp. 9181–9185.

References

72. Levitt, J. A., Chung, P.-H., Kuimova, M. K., Yahioglu, G., Wang, Y., Qu, J., and Suhling, K. (2011) *ChemPhysChem*, **12**, pp. 662–672.
73. Zhou, C., Yuan, L., Yuan, Z., Doyle, N. K., Dilbeck, T., Bahadur, D., Ramakrishnan, S., Dearden, A., Huang, C., and Ma, B. (2016) *Inorg. Chem.*, **55**, pp. 8564–8569.
74. Matsumoto, T., Takamine, H., Tanaka, K., and Chujo, Y. (2017) *Mater. Chem. Front.*, **1**, pp. 2368–2375.
75. Lin, Y.-C., Lin, C.-H., Chen, C.-Y., Sun, S.-S., and Pal, B. (2011) *Org. Biomol. Chem.*, **9**, pp. 4507–4517.
76. Shukla D. and Wan, P. (1993) *J. Am. Chem. Soc.*, **115**, pp. 2990–2991.
77. Angulo, G., Brucka, M., Gerecke, M., Grampp, G., Jeannerat, D., Milkiewicz, J., Mitrev, Y., Radzewicz, C., Rosspeintner, A., Vauthey, E., and Wnuk, P. (2016) *Phys. Chem. Chem. Phys.*, **18**, pp. 18460–18469.
78. Förster T. and Hoffmann, G. Z. (1971) *Phys. Chem.*, **75**, pp. 63–69.
79. Chen, J., Kistemaker, J. C. M., Robertus, J., and Feringa, B. L. (2014) *J. Am. Chem. Soc.*, **136**, pp. 14924–14932.
80. Ebe, K., Seno, H., and Horigome, K. (2003) *J. Appl. Polym. Sci.*, **90**, pp. 436–441.
81. Uchida, K., Izumi, N., Sukata, S., Kojima, Y., Nakamura, S., and Irie, M. (2006) *Angew. Chem. Int. Ed.*, **45**, pp. 6470–6473.
82. Norikane, Y., Hirai, Y., and Yoshida, M. (2011) *Chem. Commun.*, **47**, pp. 1770–1772.
83. Akiyama, H. and Yoshida, M. (2012) *Adv. Mater.*, **24**, pp. 2353–2356.
84. Hoshino, M., Uchida, E., Norikane, Y., Azumi, R., Nozawa, S., Tomita, A., Sato, T., Adachi, S., and Koshihara, S. (2014) *J. Am. Chem. Soc.*, **136**, pp. 9158–9164.
85. Uchida, E., Azumi, R., and Norikane, Y. (2015) *Nat. Commun.*, **6**, pp. 7310.
86. Shimamura, A., Priimagi, A., Mamiya, J., Ikeda, T., Yu, Y., Barrett, C. J., and Shishido, A. (2011) *ACS Appl. Mater. Interfaces*, **3**, pp. 4190–4196.
87. Takeshima, T., Liao, W., Nagashima, Y., Beppu, K., Hara, M., Nagano, S., and Seki, T. (2015) *Macromolecules*, **48**, pp. 6378–6384.
88. Zhou, H., Xue, C., Weis, P., Suzuki, Y., Huang, S., Koynov, K., Auernhammer, G. K., Berger, R., Butt, H.-J., and Wu, S. (2017) *Nature Chem.*, **9**, pp. 145–151.
89. Boyne, J. M., Millan, E. J., and Webster, I. (2001) *Int. J. Adhes. Adhes.*, **21**, pp. 49–53.

90. Trenor, S. R., Long, T. E., and Love, B. J. (2005) *J. Adhes.*, **81**, pp. 213–229.
91. Inui, T., Sato, E., and Matsumoto, A. (2012) *ACS Appl. Mater. Interfaces*, **4**, pp. 2124–2132.
92. Yamaguchi, H., Kobayashi, Y., Kobayashi, R., Takashima, Y., Hashidzume, A., and Harada, A. (2012) *Nature Comm.*, **3**, pp. 603.
93. Asadirad, A. M., Boutault, S., Erno, Z., and Branda, N. R. (2014) *J. Am. Chem. Soc.*, **136**, pp. 3024–3027.
94. Heinzmann, C., Coulibaly, S., Roulin, A., Fiore, G. L., and Weder, C. (2014) *ACS Appl. Mater. Interfaces*, **6**, pp. 4713–4719.
95. Goodby, J. W., Collings, P. J., Kato, T., Tschierske, C., Gleeson, H. F., and Raynes, P. (Eds.) (2014) *Handbook of Liquid Crystals* (Wiley-VCH).
96. Akiyama, H., Kanazawa, S., Okuyama, Y., Yoshida, M., Kihara, H., Nagai, H., Norikane, Y., and Azumi, R. (2014) *ACS Appl. Mater. Interfaces*, **6**, pp. 7933–7941.
97. Silva, L. F. M., Öchsner, A., and Adams, R. D. (Eds.) (2011) *Handbook of Adhesion Technology* (Springer-Verlag).
98. Acree, W. E. Jr. and Chickos, J. S. (2006) *J. Phys. Chem. Ref. Data.*, **35**, pp. 1051–1330.
99. Okuda, M. and Katayama, K. (2008) *J. Phys. Chem. A*, **112**, pp. 4545–4549.
100. Bhola, R., Bhola, R., Payamyar, P., Murray, D. J., Kumar, B., Teator, A. J., Schmidt, M. U., Hammer, S. M., Saha, A., Sakamoto, J., Schlüter, A. D., and King, B. T. (2013) *J. Am. Chem. Soc.*, **135**, pp. 14134–14141.
101. Kamperman, M. and Synytska, A. (2012) *J. Mater. Chem.*, **22**, pp. 19390–19401.

Chapter 3

Optical Microresonators from π-Conjugated Polymers

Yohei Yamamoto

Faculty of Pure and Applied Sciences, University of Tsukuba,
1-1-1 Tennodai, Tsukuba, Ibaraki 305-8573, Japan
yamamoto@ims.tsukuba.ac.jp

Optical microresonators play important roles for micrometer-scale laser sources, optical circuits, chemical- and bio-sensing tools, and so on. Especially, microresonators from organic and polymer materials are advantageous for their inherent soft and elastic propensities and simple fabrication process at around room temperature with low energy consumption. In this chapter, self-assembled microspheres from π-conjugated polymers are described, which act as fluorescent microresonators. Upon focused laser illumination, whispering gallery modes (WGMs) are excited, where the confined photons inside a microsphere self-interferes. As a result, the photoluminescence (PL) involves sharp and periodic PL lines. Furthermore, strong photoexcitation leads to WGM lasing from a single conjugated polymer microspheres. Because π-conjugated

---

Light-Active Functional Organic Materials
Edited by Hiroko Yamada and Shiki Yagai
Copyright © 2019 Jenny Stanford Publishing Pte. Ltd.
ISBN 978-981-4800-15-0 (Hardcover), 978-0-429-44853-9 (eBook)
www.jennystanford.com

polymers have advantages such as high photoabsorptivity, high refractive index, luminescent properties, and charge transport properties, the conjugated polymer microresonators will be a new type of resonators for optical and optoelectronic applications.

3.1 Introduction

π-Conjugated polymers, also called conducting polymers, play pivotal roles in the field of organic electronics, optoelectronics, and photonics. π-Conjugated polymers were first discovered in 1976 as a result of international, interdisciplinary research collaboration by Prof. Hideki Shirakawa (polymer chemist, Japan), Prof. Alan G. MacDiarmid (inorganic chemist, USA), and Prof. Alan J. Heeger (theoretician, USA) [1]. After the outstanding discovery, tremendous attentions have been focused on the conducting polymers, and various studies on synthesis, characterizations, fundamental properties, and applications are carried out [2]. Representative applications are solar cells, light-emitting diodes, transistors, batteries, electrochemical cells, thermoelectronics, actuators, etc., and new research fields such as soft electronics, flexible electronics, and printable electronics have been developed.

Optical resonators confine light inside the cavity [3]. The confined light interferes by itself and resonates to show up sharp lines. Micrometer-scale optical resonators are expected for applications as a very small light and laser sources, micro-optical circuits, and highly sensitive chemical and biological sensors. Especially, organic/polymer materials attract attention as new-generation optical and electronic materials, since particular microstructures such as spheres, disks, rods, tubes, and rings can be prepared by simple and low energy consumption bottom-up solution processes under ambient condition. These microstructures are useful for application in optical resonators [4]. So far, various optical microresonators consisting of small molecules, polymers, liquids, and liquid crystals have been reported, and, in many cases, a small amount of fluorescent dyes is doped in the non-fluorescent medium [5–7].

Through self-assembly studies of conjugated polymers, the authors found that certain sorts of conjugated polymers tend to form well-defined microspheres through simple solution process [8]. The microspheres exhibit WGM PL upon focused laser illumination on a single microsphere, where sharp and periodic emission lines appear in the PL spectrum caused by a light confinement and its interference inside the microsphere. Using the conjugated polymer microspheres, several studies have been carried out: energy transfer inside a sphere and between coupled spheres, laser oscillation upon femto-second pumping, as well as detailed studies on how polymers form microsphere geometry. This chapter presents an overview of the studies on conjugated polymer microresonators.

3.2 Self-Assembled Microspheres from π-Conjugated Polymers

Here, three self-assembly methods in solution process are introduced: vapor diffusion method, interface precipitation method, and miniemulsion method. In the vapor diffusion method (Fig. 3.1a), typically, a 5 mL vial containing a $CHCl_3$ solution of polymers (0.5 mg/mL, 2 mL) is placed in a 50 mL vial containing 5 mL of a nonsolvent such as MeOH. The outside vial is capped and then allowed to stand for 3 days at 25°C. The vapor of the nonsolvent is slowly diffused into the solution, resulting in a precipitation of the polymers through the supersaturated state. In the interface precipitation method (Fig. 3.1b), typically, a THF solution of a mixture of polystyrene (**PS**, [**PS**] = 1.0 mg/mL) and fluorescent dye ([dye] = 0.002–1.0 mg/mL) is carefully added on a nonsolvent layer of a water or water/EtOH mixture (1 mL). A slow diffusion of the solvents, along with simultaneous evaporation of THF to air, results in a precipitation after 6 h. In the miniemulsion method (Fig. 3.1c), a $CHCl_3$ solution of polymers is added to an aqueous solution of sodium *n*-dodecyl sulfate (SDS), and the water/$CHCl_3$ two-phase-separated solution is emulsified by vigorously stirring with homogenizer. The resultant emulsion is allowed to stand for 24 h at 25°C in an atmosphere to naturally evaporate $CHCl_3$. The excess SDS is removed by exchanging the supernatant water through centrifugation (thrice) to obtain a precipitate of conjugated polymers.

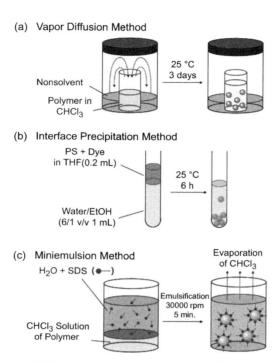

Figure 3.1 (a–c) Schematic representations of self-assembly method in solution; vapor diffusion method (a), interface precipitation method (b), and miniemulsion method (c).

Before 2010, reports on conjugated polymer microspheres were limited [9]. Later, several techniques were developed to yield spheres in the scale of sub-micrometer to several micrometers. For example, Yabu et al. reported self-assembled, sub-micrometer-scale spheres by a self-organization precipitation (SORP) method [10]. Kuehne et al. developed monodisperse conjugated polymer particles that form colloidal crystal with selected reflection and lasing [11, 12]. One reason why π-conjugated polymers do not tend to form sphere geometry is referred to the high crystallinity due to the coplanar main chain. During the precipitation process from a solution, polymers with rigid and coplanar backbone tend to stack on one another to form a crystalline domain, which grows anisotropically to a certain direction. Because a sphere has an isotropic geometry, the anisotropic growth of the π-conjugated polymers hardly yields spherical geometry in a thermodynamic process but forms one-dimensional fibers, two-dimensional lamellae structures, or irregular aggregates. Accordingly, one strategy to obtain a spherical geometry by a precipitation method is to reduce the crystallinity of the π-conjugated polymer backbone.

To prevent the π–π stacking of the conjugated planes, introduction of bulky moieties onto the main chain of the π-conjugated backbone is effective. For example, a typical semiconducting polymer F8T2, comprising dioctylfluorene and bithiophene repeating unit, hardly forms a sphere geometry upon diffusion of a nonsolvent vapor into the solution but only give irregular aggregates (Fig. 3.2a). However, introduction of four methyl groups onto the 3- and 4-positions of the thiophene rings (F8TMT2) results in a precipitation of well-defined microsphere geometry under identical self-assembly condition (Fig. 3.2b) [13]. The density functional theory (DFT) calculation clearly shows that the bithiophene moiety is largely twisted by the introduction of the methyl groups [14]. The dihedral angle between the neighboring thiophene rings in F8TMT2 is 66.5° (Fig. 3.2c), which is far larger than that in F8T2 (~5.3°, Fig. 3.2d). Such large twisting of the π-conjugated planes in F8TMT2 inhibits interpolymer π–π stacking, leading to an amorphous aggregation. Because these polymers are highly hydrophobic, the slow diffusion of the polar nonsolvents results in a nucleation and growth of the polymers while minimizing the surface area, thus leading to a formation of spheres. These two factors—low crystallinity caused by a large twisting angle or bulky side chains and slow diffusion of polar nonsolvent—are essential for the formation of microspheres from π-conjugated polymers during the precipitation process.

Following these rules, we can explain the self-assembling behaviors of conjugated polymers (Fig. 3.3). For example, highly crystalline π-conjugated polymers such as regioregular (RR)-P3HT and polyfluorene (F8) hardly form spherical geometry. Poly-*para*-phenylenevinylene (PPV) derivatives partly form microspheres [15]. Also, highly linear polymers with rigid and coplanar π-conjugated main chains such as F8T2, F8BT, and PCPDTBT, hardly form sphere geometry. In contrast, π-conjugated polymers with bulky moieties on the side chain such as F8EDOT, ArTMT2 (Ar = F8, 2,7-Cz, 3,6-Cz, DOP, PT), AZOANI, and PhTBT form microspheres by the vapor diffusion process [16, 17]. One exception is F8TPD, which are expected to form highly coplanar structure according to the DFT calculation [18]. A careful observation of the assembling process during the diffusion of MeOH into the $CHCl_3$ solution indicates that F8TPD undergoes two-step assembly; at the volume fraction of $CHCl_3$ (f_{CHCl_3}) of ~0.85, F8TPD transforms from linear to folding configuration, and further diffusion of MeOH at f_{CHCl_3} of 0.72 results in an interpolymer assembly to form microspheres.

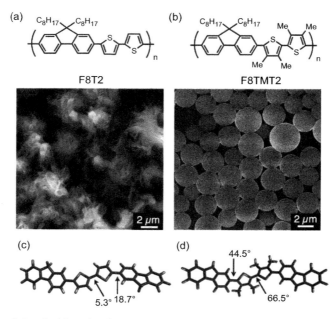

Figure 3.2 (a, b) Molecular structures of **F8T2** (a) and **F8TMT2** (b) and SEM images of the resultant precipitates obtained by vapor diffusion method. (c, d) Optimized molecular structures of fluorene–bithiophene–fluorene trimer (c) and fluorene–tetramethylbithiophene–fluorene trimer (d) by DFT calculation. The twisting angles between the neighboring rings are denoted.

Later, we found that other methods such as interface precipitation and miniemulsion methods also produce microspheres effectively. Especially, by the miniemulsion method, polymers with high crystallinity can give microspheres [19]. However, these methods afford microspheres as a kinetic product; thereby in some cases, the sphericity and surface smoothness are worse than the microspheres produced by the vapor diffusion method. In addition, in the miniemulsion method, surfactant covers the microspheres; hence a thorough washing of the microspheres after precipitation is necessary.

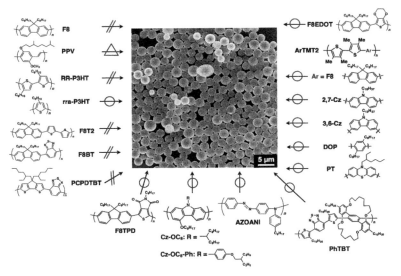

Figure 3.3 Summary of conjugated polymers that form microsphere geometry (circle) or not (double slashes) by the vapor diffusion method. Triangle; microsphere forms in part.

3.3 WGM Photoluminescence from Conjugated Polymer Microspheres

The conjugated polymer microspheres act as an optical resonator. Upon excitation of a single microsphere with a focused laser beam, PL spectrum from a single sphere involves sharp and periodic PL lines superimposed on a broad PL band originating from PL of the polymer (Fig. 3.4a) [16]. These characteristic PL lines are derived from WGMs. The PL generated inside the microsphere is confined via total internal reflection (TIR) at the polymer/air interface and interferes with round propagation (Fig. 3.4b).

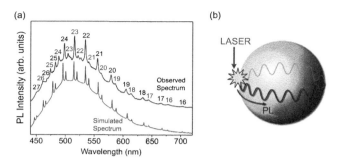

Figure 3.4 (a) PL spectrum from a single microsphere of **F8TMT2**, upon focused laser excitation at 405 nm. Black and gray numbers indicate TE and TM modes, respectively. The bottom curve shows simulated spectrum. (b) Schematic representation of the confinement of PL in a sphere.

The first-order resonant condition is described as

$$n\pi d = l\lambda \quad (3.1)$$

where n is the refractive index, d is the diameter, l is an integer, and λ is wavelength, and resonance occurs when the one-circle optical path length coincides with the integer multiple of the wavelength of PL. Because the transverse electric (TE) and transverse magnetic (TM) modes split in WGM, the resonant wavelengths of each mode are described as the simplified Oraevsky's Eqs. (3.2) and (3.3) [20]:

$$\lambda_l^E = 2\pi r(\varepsilon\mu)^{\frac{1}{2}} \left[l + \frac{1}{2} + 1.85576\left(l+\frac{1}{2}\right)^{\frac{1}{3}} - \frac{1}{\varepsilon}\left(\frac{\varepsilon\mu}{\varepsilon\mu-1}\right)^{\frac{1}{2}} \right]^{-1} \quad (3.2)$$

$$\lambda_l^H = 2\pi r(\varepsilon\mu)^{\frac{1}{2}} \left[l + \frac{1}{2} + 1.85576\left(l+\frac{1}{2}\right)^{\frac{1}{3}} - \frac{1}{\mu}\left(\frac{\varepsilon\mu}{\varepsilon\mu-1}\right)^{\frac{1}{2}} \right]^{-1} \quad (3.3)$$

where λ_l^E and λ_l^H are the wavelengths of the l-th TE and TM modes, respectively; ε $(=n^2)$ is the dielectric permittivity; μ $(=1)$ is the magnetic permeability, and r is the sphere's radius $(=d/2)$.

The WGM PL of conjugated polymer microsphere was first observed from **F8TMT2** [16]. Figure 3.4a shows the PL spectrum from a single microsphere of **F8TMT2**. Theoretical simulation using Eqs. (3.2) and (3.3) matches well with the resonant PL peaks observed experimentally (Fig. 3.4a, bottom). WGMs were further observed from various conjugated polymer spheres such as **2,7-CzTMT2**, **PTTMT2**, **AZOANI** [16], isolated conjugated polymer **PhTBT** [17], the poly-*para*-phenylenevinylene derivative **PPV** [15], fluorene copolymer **F8TPD** [21, 22], the carbazole polymers **PBDTCCz** [21], **Cz-OC$_8$**, and **Cz-OC$_8$-Ph** [23].

The WGM is hardly observed when d is less than 2 μm, because the microspheres with small d have a large curvature that lowers the efficiency of the TIR at the polymer/atmosphere (air) interface. When d is larger than 2 μm, clear periodical WGM lines appear in the PL spectrum, and, as d is larger, the WGM lines are much more crowded because the one-circle optical path involves a larger number of waves of light [16]. The Q-factor—defined by the peak wavelength divided by the full width at half maximum of the peak— of a microsphere of **F8TPD** reaches as high as 2200 as a result of its high sphericity and superior PL quantum yield [21]. WGM is quite sensitive to the shape and surface morphology of the microspheres; in the case of self-assembly of the isolated conjugated polymer **PhTBT**, microspheres form when $CHCl_3$ and CH_2Cl_2 are used as good solvents. However, the microspheres prepared from CH_2Cl_2 exhibit WGM PL only because of their high degree of sphericity [17].

Microspheres of **Cz-OC$_8$** and **Cz-OC$_8$-Ph** exhibit white-color WGM PL [23]. Upon laser excitation at 405 nm, periodic WGM PL peaks appear in a whole visible-to-NIR region possibly because of the partial oxidation of the carbazole group. The Commission Internationale de L'éclairage (CIE) coordinates shift from (0.17, 0.11) to (0.40, 0.42) upon photoirradiation. The high brightness resonant photoemitter is valuable for micrometer-scale, white-color light source and will further be applied as full-color light sources by sorting the necessary emission lines.

The conjugated polymer microspheres can be wrapped by graphene oxide (GO) nanosheets when the suspension of the microspheres is added to an aqueous solution of GO [24]. Because GO is water dispersible, the resultant GO-wrapped polymer microspheres are highly dispersive in water. Due to the deterioration of the surface smoothness, Q-factors of the microcavity decrease by half in comparison with the microspheres without GO wrapping.

Optically induced mode splitting is observed from high Q-factor, conjugated polymer microcavities [22]. Upon successive excitation in air, a single WGM (Q = 6800) splits into a series of fine peaks with Q as high as 10,000. The strong photoirradiation causes partial destruction of the π-conjugated systems, leading to a prolate distortion of the optical shape, that is, the change of the refractive index distribution, which split the degenerated azimuthal mode of WGM.

3.4 Inter-sphere Energy Transfer Cascade through Coupled Microspheres

Excitation energy transfer behaviors inside and through microcavities are studied with energy-donating F8TPD and energy-accepting PBDTCCz [21]. As mentioned in Sec. 3.3., both polymers self-assemble to form microspheres under similar vapor diffusion condition. Then, F8TPD with three different molecular weights (P1$_{16k}$, P1$_{43k}$, and P1$_{80k}$ with M_n = 16.2, 43.0, and 79.6 kg/mol, respectively) and PBDTCCz with M_n = 11.0 kg/mol (P2) are prepared, and the self-assembly behaviors of the mixtures of these polymers are studied (Fig. 3.5). In the case of a mixture of P1$_{16k}$/P2 with similar molecular weights, microspheres with average diameters of ~3 μm are obtained reproducibly from all mixing ratio investigated. On the other hand, only ill-defined aggregates are produced with combinations with largely different molecular weights such as P1$_{43k}$/P2 and P1$_{80k}$/P2. PL spectra of the cast films of the microspheres with excitation wavelength (λ_{ex}) of 470 nm show that PLs are largely red-shifted with only a small weight fraction of P2 (f_{P2}), indicating that excitation energy transfer takes place efficiently from P1$_{16k}$ to P2 inside the microspheres. Interestingly, the PL quantum yield (ϕ_{PL}) of P1$_{16k}$/P2 blend microspheres is enhanced compared to those made from homotropic assembly of P1$_{16k}$ or P2 due to the excitation energy transfer and subsequent exciton localization in P2, highly dispersed in the P1 medium.

The μ-PL measurements of single microspheres of P1$_{16k}$/P2 display that each microsphere shows WGM PL, and the wavelength range of WGM shifts from 550–650 nm to 600–750 nm because of energy transfer from P1$_{16k}$ to P2 inside the spheres (Fig. 3.6a–c). From the fluorescent microscopy images, clear color change from yellow (PL from P1$_{16k}$) to red (PL from P2) are observed. Furthermore, when the spheres of P1$_{16k}$ and P2 are coupled and focused laser is irradiated to a perimeter of the sphere of P1$_{16k}$, inter-sphere energy transfer takes place at the contact point of the bisphere, and red-colored PL is observed at the opposite side of the P2 sphere (Fig. 3.6d,e). The transfer efficiency is much higher when P1$_{16k}$/P2 (f_{P2} = 0.2) is used instead of the P2 sphere, possibly because the light confinement property of the polymer blend sphere is superior to that of the P2 sphere (Fig. 3.6f,g).

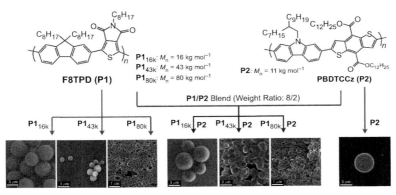

Figure 3.5 Molecular structures of **F8TPD** (**P1**) and **PBDTCCz** (**P2**) and SEM images of the self-assembled precipitates upon MeOH vapor diffusion into a CHCl$_3$ solution of **P1** and/or **P2**.

In order to thoroughly investigate the cavity-mediated energy transfer, we prepared PS microspheres, doped with a fluorescent dye, BODIPY (Fig. 3.7a) [25]. BODIPY displays various fluorescent colors depending on the assembling manners. By an interface precipitation method (Fig. 3.1b), PS forms a microsphere, and the PL color changes from green, yellow, orange, and red, by increasing the initial concentration of BODIPY (Fig. 3.7b–e). Each microsphere displays clear WGM PL at its PL wavelength region (Fig. 3.7f). When two spheres are coupled and the perimeter of one sphere is excited, energy transfer occurs through the contact point if the spheres satisfy the energy donor and acceptor combination (Fig. 3.7h,g). The relative energy transfer efficiency approximately follows fluorescence resonance energy transfer rule; it is proportional to the product of the spectral overlap between the PL band of the energy donor and the photoabsorption band of the energy acceptor and fluorescence quantum yield. Therefore, the energy transfer is directional, which reaches as long as several micrometer distances due to the confinement of the excited state. In fact, when three spheres with green, green, and orange PL colors are connected linearly, and the center sphere is photoexcited, energy transfer mainly occurs only toward the right direction (green → orange), and the left green-emissive sphere hardly fluoresces (Fig. 3.7h). On the other hand, when the right orange-emissive sphere is photoexcited, energy transfer hardly occurs to the center, green-emissive sphere (Fig. 3.7i). Furthermore, energy transfer cascade is observed when three microspheres with the fluorescence color of green, yellow, and orange are connected linearly on this order and green fluorescent sphere is excited.

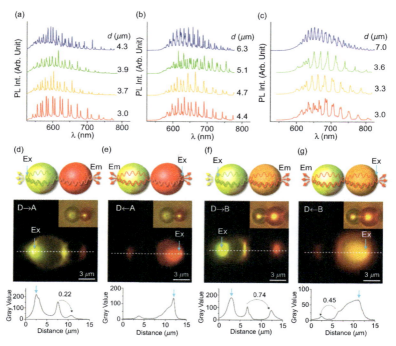

Figure 3.6 (a–c) PL spectra of a single microsphere of **P1** (a), **P1/P2** blend (b), and **P2** (c) upon focused laser excitation at 470 nm at the perimeter of the microsphere. The numerals indicate diameter of the microsphere in micrometer. (d–g) Schematic representations of coupled bispheres of **P1–P2** (a and b) and **P1–P1/P2** blend (c and d), fluorescence microscopy images upon focused laser irradiation at the left or right part of the bispheres, and cross section profiles of the gray value of the images (dotted line).

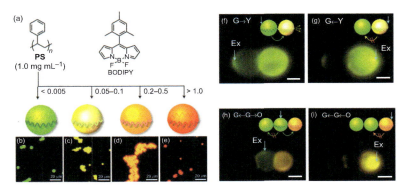

Figure 3.7 (a) Chemical structures of **PS** and **BODIPY**, and schematic representations of the self-assembled microspheres with different fluorescent colors. The values indicate initial concentration of **BODIPY** in mg mL^{-1}. (b–e) Fluorescence microscopy images of the resultant microspheres. (f, g) Fluorescence microscopy images of bisphere with green and yellow PL colors upon excitation at the left and right edge of the bisphere. (g, h) Fluorescence microscopy images of trisphere with green, green, and orange PL colors upon excitation at the center and right edge of the trisphere. Scale bars: 3 μm.

3.5 WGM Lasing from Conjugated Polymer Microspheres

To obtain WGM lasing from the microspheres, population inversion and stimulated emission are necessary. For population inversion, strong optical pumping is needed, so polymers have to be tough against strong photoirradiation. In addition, excited-state absorption causes a loss of the stimulated emission, which suppresses the optical gain. To satisfy these conditions, **F8**, **F8BT**, and **PPV** are suitable for lasing. However, these polymers hardly form microspheres by the vapor diffusion method because of the high crystallinity, which prevent amorphous assembly as the thermodynamic product.

In order to obtain microspheres from conjugated polymers with high crystallinity, miniemulsion method is adopted (Fig. 3.1c) [19]. By the miniemulsion method, **F8**, **F8BT**, and **PPV** form microsphere (Fig. 3.8a,d,g). Upon femto-second laser pumping (Δ = 300 fs, λ_{ex} = 397 nm, f = 1 kHz), a single microsphere of **F8** displays sharp peaks at around 463 nm, typical of laser oscillation when the pumping fluence increases above 1.5 µJ/cm² (Fig. 3.8b,c). Microspheres of **F8BT** and **PPV** also display lasing behaviors when the pumping fluence is beyond 58 and 25 µJ/cm, respectively (Fig. 3.8e,f,h,i). The lasing threshold is more than one order of magnitude higher than that of **F8**, possibly because of the mismatch of the absorption band with the pumping wavelength.

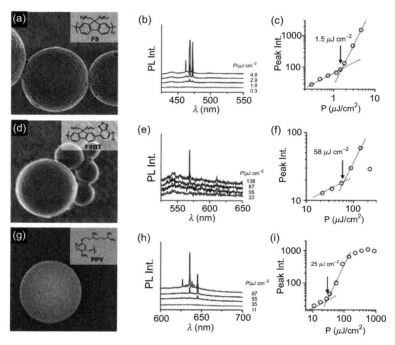

Figure 3.8 SEM micrographs of self-assembled microspheres (a, d, g), PL spectra upon femto-second pumping (b, e, h), and plots of the PL peak intensity versus power density (c, f, i) of **F8** (a), **F8BT** (d), and **PPV** (g).

The lasing threshold is reduced by one-fourth when the microsphere is immobilized on an Ag-coated substrate (thickness of the Ag layer: 30 nm). Finite-difference time-domain (FDTD) simulations revealed that the enhanced PL intensity with reduced lasing threshold could be mainly due to a mirror effect of the Ag layer, in addition to a minor contribution of the near-field enhancement.

3.6 Summary and Prospects

This chapter describes self-assembly of π-conjugated polymers to form well-defined microspheres. The important factors for the formation of microspheres are (1) low crystallinity of the polymer chain and (2) slow diffusion of polar nonsolvent into the solution of the π-conjugated polymers. The size and distribution of the microspheres are generally controlled by the self-assembly condition; slow precipitation results in large microspheres with broad size distribution. Upon focused laser irradiation on a single microsphere, the conjugated polymer microspheres exhibit WGMs, involving sharp and periodic PL lines due to the interference of the confined photons inside the microsphere. The microspheres show various optical properties, such as intra- and inter-sphere energy transfers, optically induced mode splitting, and laser oscillation.

Conjugated polymer microresonators are valuable with respect to the following points: (1) high refractive index (1.6–2.0), (2) high absorptivity and PL efficiency, and (3) simple and low-cost fabrication. For the practical application of microspheres, all of these factors are important. Future prospects are electrically driven WGM PL and lasing by charge injection into conjugated polymer microresonators. For practical applications, a microsphere array on a substrate or electrode surface is necessary.

Acknowledgments

The author acknowledges Prof. Takaki Kanbara, Prof. Masashi Kijima, Prof. Junpei Kuwabara, Prof. Tatsuya Nabeshima, and Prof. Takashi Nakamura in the University of Tsukuba; Dr. Chengjun Pan, Dr. Kazunori Sugiyasu, and Dr. Masayuki Takeuchi in the National Institute for Materials Science (NIMS) for the syntheses of conjugated polymers and fluorescent dyes; Prof. Axel Lorke and Dr. Daniel Braam in the University of Duisburg-Essen; Dr. Thang Dao, Dr. Satoshi Ishii, and Dr. Tadaaki Nagao in the NIMS for μ-PL experiments; Dr. Fumio Sasaki in the National Institute for Advanced Industrial Science and Technology (AIST) for femto-second laser pumping experiments; Dr. Jer-Shing Huang in the Leibniz Institute for Optical Technology (IPHT) in Jena for FDTD simulations. The author also acknowledges students in the University of Tsukuba, Ms. Taeko Adachi, Mr. Kenichi

Tabata, Mr. Tong Liang, Mr. Soh Kushida, Mr. Daichi Okada, and Mr. Yusuke Aikyo, for developing this research.

This work was supported by KAKENHI (25708020, 15H00860, 15H00986, 15K13812, 16H02081, 17H05142) from JSPS/MEXT, Japan, Asahi Glass Foundation, University of Tsukuba-DAAD partnership program, University of Tsukuba pre-strategic initiative, and TIA-KAKEHASHI.

References

1. Shirakawa, H., Louis, E. J., MacDiarmid, A. G., Chiang, C. K., and Heeger, A. J. (1977). Synthesis of electrically conducting organic polymers: Halogen derivatives of polyacetylene, (CH)$_x$, *J. Chem. Soc. Chem. Commun.*, pp. 578–580.

2. Cicoira, F. and Santato, C. (2013). *Organic Electronics: Emerging Concepts and Technologies* (WILEY-VCH, Germany).

3. Vahala, K. J. (2003). Optical microcavities, *Nature*, **424**, pp. 839–846.

4. Zhang, W., Yao, J., and Zhao, Y.-S. (2016). Organic micro/nanoscale lasers, *Acc. Chem. Res.*, **49**, pp. 1691–1700.

5. Tang, S. K. Y., Derda, R., Quan, Q., Lončar, M., and Whitesides, G. M. (2011). Continuously tunable microdroplet-laser in a microfluidic channel, *Opt. Express*, **19**, pp. 2204–2215.

6. Humar, M., Ravnik, M., Pajk, S., and Musevic, I. (2009). Electrically tunable liquid crystal optical microresonators, *Nat. Photon.*, **3**, pp. 595–600.

7. Ta, V. D., Chen, R., and Sun, H. D. (2013). Tuning whispering gallery mode lasing from self-assembled polymer droplets, *Sci. Rep.*, **3**, 1362.

8. Yamamoto, Y. (2016). Spherical resonators from π-conjugated polymers, *Polym. J.*, **48**, pp. 1045–1050.

9. Pecher, J. and Mecking, S. (2010). Nanoparticles of conjugated polymers, *Chem. Rev.*, **110**, pp. 6260–6279.

10. Yabu, H. (2012). Creation of functional and structured polymer particles by self-organized precipitation (SORP), *Chem. Lett.*, **85**, pp. 265–274.

11. Kuehne, A. J. C., Gather, M. C., and Sprakel, J. (2012). Monodisperse conjugated polymer particles by Suzuki–Miyaura dispersion polymerization, *Nat. Commun.*, **3**, 1088.

12. Mikosch, A., Ciftci, S., and Kuehne, A. J. C. (2016). Colloidal crystal lasers from monodisperse conjugated polymer particles via bottom-up coassembly in a sol–gel matrix, *ACS Nano*, **10**, pp. 10195–10201.
13. Adachi, T., Tong, L., Kuwabara, J., Kanbara, T., Saeki, A., Seki, S., and Yamamoto, Y. (2013). Spherical assemblies from π-conjugated alternating copolymers: Toward optoelectronic colloidal crystals, *J. Am. Chem. Soc.*, **135**, pp. 870–876.
14. Tong, L., Kushida, S., Kuwabara, J., Kanbara, T., Ishii, N., Saeki, A., Seki, S., Furumi, S., and Yamamoto, Y. (2014). Tetramethylbithiophene in π-conjugated alternating copolymers as an effective structural component for the formation of spherical assemblies, *Polym. Chem.*, **5**, pp. 3583–3587.
15. Kushida, S., Braam, D., Lorke, A., and Yamamoto, Y. (2015). Whispering gallery mode photoemission from self-assembled poly-*para*-phenylenevinylene microspheres, *AIP Conf. Proc.*, **1702**, 090046.
16. Tabata, K., Braam, D., Kushida, S., Tong, L., Kuwabara, J., Kanbara, T., Beckel, A., Lorke, A., and Yamamoto, Y. (2014). Self-assembled conjugated polymer spheres as fluorescent microresonators, *Sci. Rep.*, **4**, 5902.
17. Kushida, S., Braam, D., Pan, C., Dao, T. D., Tabata, K., Sugiyasu, K., Takeuchi, M., Ishii, S., Nagao, T., Lorke, A., and Yamamoto, Y. (2015). Whispering gallery resonance from self-assembled microspheres of highly fluorescent isolated conjugated polymers, *Macromolecules*, **48**, pp. 3928–3933.
18. Kushida, S., Oki, O., Saito, H., Kuwabara, J., Kanbara, T., Tashiro, M., Katouda, M., Imamura, Y., and Yamamoto, Y. (2017). From linear to foldamer and assembly: Hierarchical transformation of a coplanar conjugated polymer into a microsphere, *J. Phys. Chem. Lett.*, **8**, pp. 4580–4586.
19. Kushida, S., Okada, D., Sasaki, F., Lin, Z.-H., Huang, J.-S., and Yamamoto, Y. (2017). Low-threshold whispering gallery mode lasing from self-assembled microspheres of single-sort conjugated polymers, *Adv. Opt. Mater.*, **5**, 1700123.
20. Oraevsky, A. N. (2002). Whispering-gallery waves, *Quant. Elect.*, **32**, pp. 377–400.
21. Kushida, S., Braam, D., Dao, T., Saito, H., Shibasaki, K., Ishii, S., Nagao, T., Saeki, A., Kuwabara, J., Kanbara, T., Kijima, M., Lorke, A., and Yamamoto, Y. (2016). Conjugated polymer blend microspheres for efficient, long-range light energy transfer, *ACS Nano*, **10**, pp. 5543–5549.

22. Braam, D., Kushida, S., Niemöller, R., Prinz, G. M., Saito, H., Kanbara, T., Kuwabara, J., Yamamoto, Y., and Lorke, A. (2016). Optically induced mode splitting in self-assembled, high quality-factor conjugated polymer microcavities, *Sci. Rep.*, **6**, 19635.
23. Kushida, S., Okabe, S., Dao, T. D., Ishii, S., Nagao, T., Saeki, A., Kijima, M., and Yamamoto, Y. (2016). Self-assembled polycarbazole microspheres as single-component, white colour resonant photoemitters, *RSC Adv.*, **6**, pp. 52854–52857.
24. Aikyo, Y., Kushida, S., Braam, D., Kuwabara, J., Kondo, T., Kanbara, T., Nakamura, J., Lorke, A., and Yamamoto, Y. (2016). Enwrapping conjugated polymer microspheres with graphene oxide nanosheets, *Chem. Lett.*, **45**, pp. 1024–1026.
25. Okada, D., Nakamura, T., Braam, D., Dao, T., Ishii, S., Nagao, T., Lorke, A., Nabeshima, T., and Yamamoto, Y. (2016). Color-tunable resonant photoluminescence and cavity-mediated multistep energy transfer cascade, *ACS Nano*, **10**, pp. 7058–7063.

Chapter 4

Hydrogen-Bond-Directed Nanostructurization of Oligothiophene Semiconductors for Organic Photovoltaics

Xu Lin,[a] Hayato Ouchi,[b] and Shiki Yagai[b,c]

[a]*Engineering Laboratory for Highly-Efficient Utilization of Biomass, College of Materials Engineering, Southwest Forestry University, 300 Bailong Road, Kunming 650224, Yunnan Province, China*
[b]*Division of Advanced Science and Engineering, Graduate School of Science and Engineering, Chiba University, 1-33 Yayoi-cho, Inage-ku, Chiba 263-8522, Japan*
[c]*Institute for Global Prominent Research (IGPR), Chiba University 1-33 Yayoi-cho, Inage-ku, Chiba, 263-8522, Japan*
yagai@faculty.chiba-u.jp

4.1 Introduction

The organization of synthetically accessible π-conjugated small molecules through programmed self-assembly processes provides specifically designed nanostructures with desirable optoelectronic properties [1, 2]. One-dimensional columnar nanostructures

---

Light-Active Functional Organic Materials
Edited by Hiroko Yamada and Shiki Yagai
Copyright © 2019 Jenny Stanford Publishing Pte. Ltd.
ISBN 978-981-4800-15-0 (Hardcover), 978-0-429-44853-9 (eBook)
www.jennystanford.com

spontaneously formed by discoid or star-shaped semiconductor molecules are reasonable nanostructures as media for the quasi-one-dimensional transport of charge carriers [3–6]. Triphenylenes, hexabenzocoronenes, and other discoid molecules that can effectively organize into such columnar nanostructures have been widely explored as solution-processable semiconductor materials [1]. They are particularly promising as materials for bulk heterojunction (BHJ) organic photovoltaics (OPV) [7–9] because BHJ-OPV devices are expected to show efficient performance if the segregated structures of donor and acceptor materials are controlled in nanoscale precision (Fig. 4.1) [10].

Without using the above extended π-conjugated molecules, well-defined columnar nanostructures can be constructed from synthetically more accessible small molecular building blocks through programmed self-assembly processes, which would be a more cost-effective strategy for the development of small molecular BHJ-OPV (Fig. 4.1). Many supramolecular chemists have engaged with the design of supramolecular building blocks that can organize into columnar nanostructures and found that the use of multiple hydrogen-bonding interactions is a very effective means to construct well-defined columnar nanostructures via the formation of supramolecular discotics, so-called supermacrocycles or rosettes [11–19]. Despite a great advance in the synthesis and design of such hierarchically organized columnar nanostructures, little attention has been focused on their application in BHJ-OPV. This situation might be largely due to the requirement of highly soluble but "electronically inactive" long aliphatic chains to ensure programmed self-assembly processes as well as to stabilize hierarchically organized columnar nanostructures. These long aliphatic chains, which in most cases cover the semiconducting columnar core, not only hamper efficient charge separation at the p–n interface but also deteriorate charge transportation between the nanostructures in BHJ films. Accordingly, properly designed hydrogen-bonding small molecules that can organize into well-organized columnar nanostructures but are free from long aliphatic chains are applicable to BHJ solar cells. In this chapter, we focus on our quest for such supramolecular BHJ-OPV materials based on multiple hydrogen-bonding interactions.

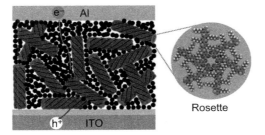

Figure 4.1 Schematic representation of the transportation of a charge carrier photogenerated at the interface between p-type semiconducting columnar nanostructures and n-type small molecular materials.

4.2 Insulated Semiconducting Supramolecular Nanorods Composed of Hydrogen-Bonded Oligothiophene Rosettes

The activity of our research group has been dedicated to the design, synthesis, and construction of π-conjugated or dye-based molecular assemblies by means of complementary multiple hydrogen-bonding interactions between triaminotriazine and cyanuric acid or triaminotriazine and barbituric acid [20–22]. During our investigation, we accidentally found that the barbituric acid unit itself could be a powerful functional group for the organization of various π-conjugated units into one-dimensional nanostructures [23]. Thus far, we have reported that a variety of nanostructures such as cylindrical fibers [23, 24], rings [24, 25], coils [25], interlocked rings (catenanes) [26], and helically folded supramolecular polymers [27] can be obtained from π-conjugated units whose one end is equipped with barbituric acid, while the other end with a tridodecyloxyphenyl aliphatic tail (e.g., **1** in Fig. 4.2a). For all of the systems, the formation of hexameric rosettes as intermediate building blocks was proposed on the basis of powder X-ray diffraction (PXRD) analyses even though the literature contains no reports regarding the formation of barbituric acid rosettes. Such a hydrogen-bond-mediated organization into well-defined nanostructures motivated us to design barbituric-acid-functionalized semiconducting small molecules that are applicable to BHJ-OPV.

We first designed and synthesized the oligothiophene derivative **1** shown in Fig. 4.2a [23]. This compound formed one-dimensional nanorods through the formation of hydrogen-bonded rosette [19, 28] in low-polarity solvents or in the bulk state. Although the nanorods showed excellent intrinsic hole mobility, as measured by flash-photolysis time-resolved microwave conductivity (FP-TRMC, see Chapter 5 by Saeki), the BHJ film fabricated from blend solutions of **1** and $PC_{61}BM$ ([6,6]-phenyl-C_{61}-butyric acid methyl ester) showed quite poor photovoltaic performance (power conversion efficiency, PCE < 0.1%). The long alkyl chains surrounding the stacked semiconducting oligothiophene moieties are most likely responsible for the poor device performance (Fig. 4.2b), because they essentially hamper the charge separation with the $PC_{61}BM$ and charge-carrier migration between nanorods.

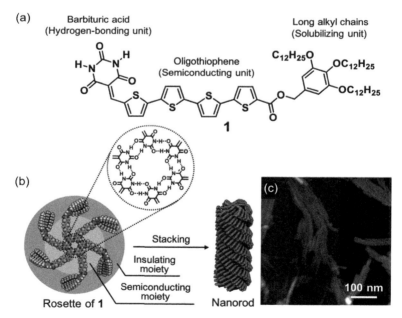

Figure 4.2 (a) Chemical structure of **1**. (b) Schematic representation of hierarchical organization of **1** into nanorods through the formation of hydrogen-bonded rosette. (c) AFM image of nanorods of **1** spin-cast from a methylcyclohexane solution onto highly oriented pyrolytic graphite (HOPG).

4.3 Semiconducting Supramolecular Nanorods Composed of Hydrogen-Bonded Oligothiophene Rosettes

An effective strategy for solving the issue that compound **1** faced was the removal of peripheral long alkyl chains covering the rod-like nanostructures while maintaining their sufficient solubility in organic solvents. For this purpose, we designed the oligothiophene derivative **2a** wherein solubilizing alkyl chains are grafted onto the π-conjugated backbone (Fig. 4.3a) [29]. If this molecule forms hydrogen-bonded rosette, the hexyl chains should be embedded inside the rosette, and the semiconducting stacks of π-conjugated

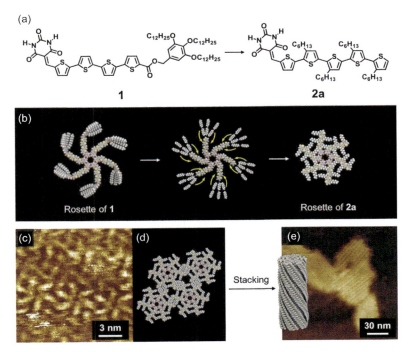

Figure 4.3 (a) Structural modification from **1** to **2a**. (b) Schematic representation of grafting alkyl chains onto oligothiophene backbones. (c) STM image of **2a** at the 1-phenyloctane–HOPG interface. (d) Proposed two-dimensional packing model of counterclockwise (CCW) rosettes of **2a**. (e) AFM image of nanorods of **2a** formed upon drop casting a toluene solution onto HOPG.

moieties should be exposed to the exterior when the rosette further stacks into rod-like nanostructure (Fig. 4.3b). The solubility of **2a** in chloroform, which is generally used for fabricating BHJ-OPV, was almost the same as that of **1**. As a serendipitous outcome of this molecular design, we obtained a clear evidence of the formation of a six-membered rosette by barbituric acid compound for the first time, by means of scanning tunneling microscopy (STM) at a liquid–solid interface (Fig. 4.3c,d). Similar to **1**, derivative **2a** gave well-defined nanorods with 4.3 nm diameter upon casting from toluene solution (Fig. 4.3e). This diameter is in good agreement with the diameter of the rosette, demonstrating the organization of the rosettes into nanorods through π–π stacking and van der Waals interactions.

4.4 Effect of Substituted Position of Alkyl Chains on the Formation of Semiconducting Supramolecular Nanorods

One may expect a similar organization process for compound **3a**, a regioisomer of **2a** (left in Fig. 4.4a) [30]. However, we have found that these regioisomers undergo completely different but high-fidelity self-assembly pathways. STM studies at a liquid–solid interface first revealed that **3a** organizes into a tape-like molecular array, which can be characterized as an open-ended linear hydrogen-bonded motif that competes with the closed rosette motif (right in Fig. 4.4a). Moreover, ^{1}H NMR, diffusion-ordered spectroscopy (DOSY), and vapor pressure osmometry (VPO) studies of **2a** and **3a** in chloroform demonstrated the presence of small discrete species and large polydispersed assemblies, respectively (left in Fig. 4.4b). Although the exact supramolecular structures could not be estimated from these solution analyses, the contrastive assembly behaviors illustrate inherently distinct aggregation propensity of the two regioisomers.

We further investigated the self-organized structures of **2a** and **3a** by the PXRD analyses of solvent-free bulk samples, which were obtained by evaporating their chloroform solutions. For **2a**, X-ray diffraction peaks in small-angle region suggested the formation of rosettes that further organize into a rectangular 2D lattice. For **3a**, the PXRD pattern could not be indexed based on usual columnar assembling motifs, and we proposed the formation of a lamellar structure composed of tape-like hydrogen-bonded supramolecular chains observed by STM (center in Fig. 4.4b).

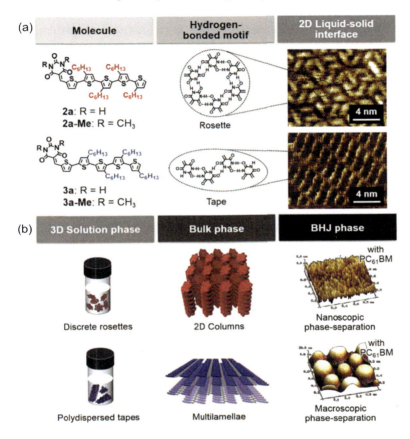

Figure 4.4 Hierarchical self-assembly of **2a** and **3a** in different phases. (a) Chemical structures of monomers (left), chemical structures of hydrogen-bonded patterns (center), and STM images at the (S)-limonene–HOPG interface. (b) Schematic representation of the hierarchical organization of **2a** and **3a** from solution (left) to bulk (center) and to BHJ films with $PC_{61}BM$ (right).

The distinct self-assembly pathways of **2a** and **3a** evolving through solvent evaporation were also manifested in a critical difference in their BHJ nanostructures with $PC_{61}BM$ (right in Fig. 4.4b). In the presence of $PC_{61}BM$ as an acceptor material, the formation of nanorods of **2a** by solution casting was hampered probably due to electronic interactions between the donor (**2a**) and the acceptor (PC$_{61}$BM) materials (Fig. 4.5a). However, thermal annealing of the resulting blend film at 80°C induced the growth of nanorods to provide a reasonably phase-separated p–n heterojunction nanostructures (Fig. 4.5b). The growth of supramolecular nanorods in the p–n blended films was reflected as the improvement in the performance of the corresponding BHJ solar cell devices from PCE = 0.8% for the as-cast film to PCE = 1.5% for the thermally annealed film at 80°C. For **2a**, further improvement in the device performance has been achieved by using PC$_{71}$BM as an acceptor, wherein PCE = 3.01% was recorded [29]. On the other hand, the device performance of **3a** and PC$_{61}$BM was very low (PCE = 0.3%) and even became lower when PC$_{71}$BM

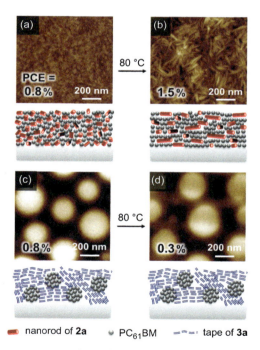

Figure 4.5 AFM images of blend films of (a, b) **2a** and PC$_{61}$BM and (c, d) **3a** and PC$_{61}$BM. (a, c) Images of as-cast films prepared from chloroform solutions. (b, d) Images obtained after thermal annealing at 80°C. Below the AFM images, schematic illustrations of the change in BHJ morphologies upon thermal annealing are shown.

was used (PCE = 0.08%). We assume that a morphological mismatch between the lamellarly organized structure of **3a** and PCBMs might be responsible for the observed macroscopic phase separation, which was confirmed by the AFM images of the blend films of **3a** and PC$_{61}$BM, which showed submicrometer-sized grains of PC$_{61}$BM (Fig. 4.5c,d). Because the above distinct structure–property relationship has not been observed for non-hydrogen-bonding references **2a-Me** and **3a-Me**, this study clearly demonstrates that small difference in molecular structures can be amplified to a huge difference in the bulk level through high-fidelity self-assembly pathways directed by directional hydrogen-bonding interactions.

4.5 Effect of π-Conjugated Length on the Formation of Semiconducting Supramolecular Nanorods

On the basis of the structure of **2a**, we also investigated the effect of the conjugation length of oligothiophene backbone on the formation of rosettes, nanorods, and also on the performance of BHJ solar cells (Fig. 4.6a). Interestingly, quaterthiophene **2b** and sextetthiophene **2c** showed STM images wherein structures of rosettes were more clearly visualized (Fig. 4.6b,d) [29]. Accordingly, we could discriminate the domains composed either of clockwise (CW) or CCW rosettes. We attributed this tendency to a difference in the efficiency of two-dimensional packing of rosettes, based on the molecular modeling (Fig. 4.6b–d, inset). The molecular modeling suggests that the rosettes of **2b** and **2c** could be two-dimensionally organized with a minimum void space, whereas **2a** leaves notable void spaces surrounded by three rosettes. Accordingly, **2a** should have fewer intermolecular and molecule–substance interactions per unit area. These results clearly demonstrate the existence of an odd or even effect in the number of thiophene unit on the two-dimensional organization of the rosettes [31]. From the perspective of columnar nanostructures composed of rosettes, this tendency should have some impact on the three-dimensional association of rod-like nanostructures.

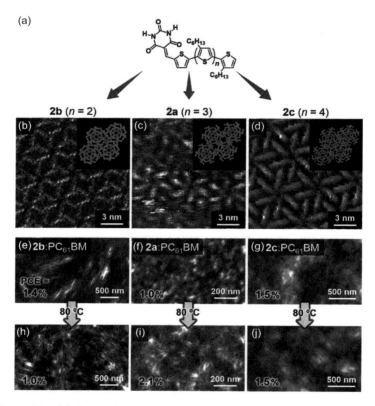

Figure 4.6 (a) Chemical structures of **2a**, **2b**, and **2c**. (b–d) STM images of (b) **2b**, (c) **2a**, and (d) **2c** at the 1-phenyloctane–HOPG interface. Only domains of CCW rosettes were shown. Inset of (b–d): Proposed two-dimensional packing models of CCW rosettes. (e–j) AFM images of the BHJ films of (e, h) **2b**, (f, i) **2a**, and (g, j) **2c** with PC$_{61}$BM from a 1:1 (v/v) mixture of chloroform and toluene, (e–g) before and (h–j) after thermal annealing at 80°C.

We thus investigated the morphologies of the BHJ films of **2b** and **2c** with PC$_{61}$BM. The AFM images of the as-cast film prepared from the mixture of **2b** and PC$_{61}$BM already showed bundled nanorods (Fig. 4.6e), which upon thermal annealing resulted in further agglomeration of nanorods (Fig. 4.6h). Reflecting the agglomeration of nanorods that should decrease the BHJ interface, the PCE of the solar cell devices dropped from 1.38% to 0.95% upon thermal annealing. For the mixture of **2c** and PC$_{61}$BM, the substantial phase separation was already suggested prior to annealing from a higher root-mean-square (RMS) roughness of the film surface, and the situation could not be improved after thermal annealing (Fig. 4.6g,j). Accordingly, the solar cell devices of **2c** and $PC_{61}BM$ did not show any improvement upon thermal annealing. The different impacts of thermal annealing on **2a** and **2b/2c** indicate that the geometries of rosettes, which influenced their two-dimensional organization, are also influential in the lateral interaction between the nanorods that promotes their agglomeration.

4.6 Functionalization of Semiconducting Supramolecular Nanorods

To improve the photovoltaic properties of our supramolecular nanorods, we synthesized oligothiophene **4** whose one end is functionalized by benzo[1,2-*b*:4,5-*b'*]dithiophene (BDT) moiety (Fig. 4.7a). BDT is a promising electron-donating unit due to its excellent charge-carrier transporting capability due to a large and planar π-conjugated structure [32]. The formation of rosettes by **4** was confirmed by STM measurements at a liquid–solid interface, although their degree of two-dimensional ordering was quite low in comparison with **2a–2c**. This is mainly due to the twisting of the π-conjugated backbone of **4**. As a result of less-ordered packing of rosettes, not only hexameric but also heptameric rosettes of **4** were observed by STM (Fig. 4.7b). The formation of supramolecular nanorods of **4** was confirmed by AFM (Fig. 4.7c). Regrettably, this compound did not show superior performance in BHJ solar cells fabricated with $PC_{61}BM$ when compared with the original molecule **2a** in terms of thermally annealed condition. However, the as-cast device of **4** with $PC_{61}BM$ showed PCE of 2.98%, which is significantly higher than that of **2a** (0.76%, Fig. 4.7d). This result was rationally explained by the results of the top-contact bottom-gate configuration organic field-effect transistor (OFET), wherein a higher charge-carrier mobility was recorded for **4** compared to **2a**. Furthermore, the maximum external quantum efficiency (EQE) of the devices of **4** was observed at 420 nm and 580 nm, which corresponds to the absorption maxima of the BDT unit and the oligothiophene backbone, respectively (Fig. 4.7e). The increased EQE response in the shorter wavelength region demonstrates the positive influence of the BDT unit both on light harvesting and the charge-carrier mobility.

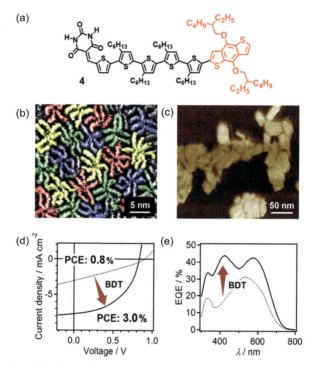

Figure 4.7 (a) Chemical structure of BDT-functionalized oligothiophene **4**. (b) STM image of **4** at the 1-phenyloctane–HOPG interface. (c) AFM image of a thin film prepared by drop casting a toluene solution of **4** onto HOPG. (d) *J–V* characteristics and (e) EQE spectra of BHJ solar cells fabricated with the as-cast BHJ films of **4** and PC$_{61}$BM (solid line) and **2a** and PC$_{61}$BM (dotted line).

4.7 Effect of Alkyl Side Chain Length on the Formation and Properties of Supramolecular Nanorods

The alkyl chains of organic semiconductors play important roles, not only in imparting enough solubility in organic solvent but also in stabilizing assembled structures. Toward understanding the relationship between molecular structures and the organization behaviors that should affect the device performance, we also investigated oligothiophene **5**, an octyl version of **2a** (Fig. 4.8a) [33]. Although the formation of rosettes by **5** was evident from STM observation, both CW and CCW rosettes were observed in randomly mixed state, which is contrastive to the cases of **2a–2c** wherein the enantiomeric rosettes separately assembled to form chiral domains (Fig. 4.8b). The lower degree of two-dimensional ordering in **5** would be due to weaker inter-rosettes interactions due to insufficient interdigitation of longer octyl chains between rosettes.

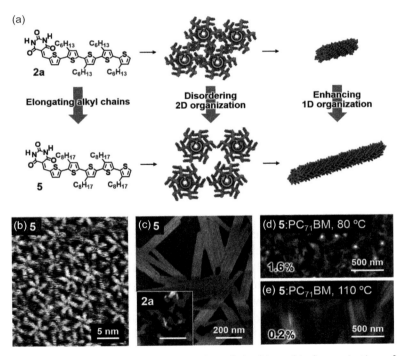

Figure 4.8 (a) Schematic representation of the hierarchical organization of **2a** and **5**. (b) STM image of **5** at the 1-phenyloctane–HOPG interface. (c) AFM image of the thin film prepared by drop casting a toluene solution of **5** onto HOPG. Inset shows AFM image of nanorod **2a** prepared by the same conditions. (d, e) AFM images of **5**:PC$_{71}$BM films after annealing at (d) 80°C and (e) 110°C.

Although the longer alkyl chains deteriorate the two-dimensional organization of rosettes, they promote the formation of rod-like nanostructures by enhancing van der Waals interaction between rosettes. Lengths of nanorods of **5** are in the range of 200–600 nm, which are considerably longer than those of **2a** (50–100 nm) fabricated under the same condition (Fig. 4.8c). For **5**, the growth of supramolecular nanorods upon thermal annealing was observed in the presence of PC$_{71}$BM, and this thermal reorganization slightly improved the device performance (PCE: 1.04 → 1.57%, Fig. 4.8d). However, upon further increasing the annealing temperature to 110°C, the nanorods further elongated to approximately 800 nm in average (Fig. 4.8e), which led to a significant decrease in the device performance (PCE: 1.57 → 0.16%). Therefore, a proper choice of alkyl chain lengths and annealing temperature should be essential to obtain supramolecular nanorods that can show sufficient photovoltaic properties.

4.8 Conclusion

In this chapter, we have reviewed a series of our researches on the design, characterization, and photovoltaic properties of supramolecular nanorods self-assembled from barbiturated oligothiophenes (Fig. 4.9). Our molecular design enabled the construction of well-defined one-dimensional nanostructures through the hydrogen-bonded supramolecular macrocycles (rosettes) without the use of long aliphatic chains. Due to the absence of exterior long aliphatic chains, BHJ solar cells fabricated with soluble fullerene derivatives exhibited performance comparable to those of the most commonly used standard polymer materials (P3HT). On one hand, an alteration of the molecular structure critically influences the self-assembly pathways, and as a result, completely different supramolecular architectures could be obtained (e.g., **2a** versus **3a**). On the other hand, careful functionalization could improve the dimensions (**5**), degree of hierarchical organization (**2b** and **2c**), and optoelectronic properties (**4**) of supramolecular nanorods without losing self-assembly pathway through the formation of rosettes. We believe that the high level of controllability in structure and property of our rosette-based semiconducting supramolecular nanorods, introduced in this chapter, can find applications not only in conventional BHJ-OPV devices but also other functional organic devices.

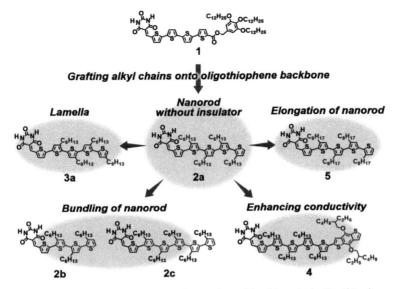

Figure 4.9 Chemical structures of a series of barbiturated oligothiophene semiconductors for BHJ solar cells.

Acknowledgments

The authors acknowledge Prof. Ken-ichi Nakayama, Prof. Takanori Fukushima, Dr. Takashi Kajitani, Prof. Shu Seki, Prof. Akinori Saeki, Prof. Hiroko Yamada, Prof. Mitsuharu Suzuki, Dr. Yoshihiro Kikkawa, Prof. Fabien Silly, Mr. Yuki Tani, Mr. Takahiro Kizaki, Ms. Mika Suzuki, Ms. Marina Gushiken, Mr. Takuya Noguchi, Mr. Tetsuro Kinoshita, and Dr. Mitsuaki Yamauchi for fruitful collaboration on the above studies. All of the above researches were supported by the Core Research for Evolutional Science and Technology (CREST), Japan Science and Technology Agency (JST).

References

1. Hoeben, F. J. M., Jonkheijm, P., Meijer, E. W., and Schenning, A. P. H. J. (2005). About supramolecular assemblies of π-conjugated systems, *Chem. Rev.*, **105**, pp. 1491–1546.
2. Wu, J., Pisula, W., and Müllen, K. (2007). Graphenes as potential material for electronics, *Chem. Rev.*, **107**, pp. 718–747.

3. Grimsdale, A. C. and Müllen, K. (2005). The chemistry of organic nanomaterials, *Angew. Chem. Int. Ed.*, **44**, pp. 5592–5629.
4. Sergeyev, S., Pisula, W., and Geerts, Y. H. (2007). Discotic liquid crystals: A new generation of organic semiconductors, *Chem. Soc. Rev.*, **36**, pp. 1902–1929.
5. Laschat, S., Baro, A., Steinke, N., Giesselmann, F., Hägele, C., Scalia, G., Judele, R., Kapatsina, E., Sauer, S., Schreivogel, A., and Tosoni, M. (2007). Discotic liquid crystals: From tailor-made synthesis to plastic electronics, *Angew. Chem. Int. Ed.*, **46**, pp. 4832–4887.
6. Kanibolotsky, A. L., Perepichka, I. F., and Skabara, P. J. (2010). Star-shaped π-conjugated oligomers and their applications in organic electronics and photonics, *Chem. Soc. Rev.*, **39**, pp. 2695–2728.
7. Thompson, B. C. and Fréchet, J. M. J. (2008). Polymer-fullerene composite solar cells, *Angew. Chem. Int. Ed.*, **47**, pp. 58–77.
8. Hains, A. W., Liang, Z., Woodhouse, M. A., and Gregg, B. A. (2010). Molecular semiconductors in organic photovoltaic cells, *Chem. Rev.*, **110**, pp. 6689–6735.
9. Mishra, A. and Bäuerle, P. (2012). Small molecule organic semiconductors on the move: Promises for future solar energy technology, *Angew. Chem. Int. Ed.*, **51**, pp. 2020–2067.
10. Yu, G., Gao, J., Hummelen, J. C., Wudl, F., and Heeger, A. J. (1995). Polymer photovoltaic cells: Enhanced efficiencies via a network of internal donor-acceptor heterojunctions, *Science*, **270**, pp. 1789–1791.
11. Kimizuka, N., Kawasaki, T., Hirata, K., and Kunitake, T. (1995). Tube-like nanostructures composed of networks of complementary hydrogen bonds, *J. Am. Chem. Soc.*, **117**, pp. 6360–6361.
12. Yang, W., Chai, X., Chi, L., Liu, X., Cao, Y., Lu, R., Jiang, Y., Tang, X., Fuchs, H., and Li, T. (1999). From achiral molecular components to chiral supermolecules and supercoil self-assembly, *Chem. Eur. J.*, **5**, pp. 1144–1149.
13. Fenniri, H., Mathivanan, P., Vidale, K. L., Sherman, D. M., Hallenga, K., Wood, K. V., and Stowell, J. G. (2001). Helical rosette nanotubes: Design, self-assembly, and characterization, *J. Am. Chem. Soc.*, **123**, pp. 3854–3855.
14. Jonkheijm, P., Miura, A., Zdanowska, M., Hoeben, F. J. M., De Feyter, S., Schenning, A. P. H. J., De Schryver, F. C., and Meijer, E. W. (2004). π-Conjugated oligo-(*p*-phenylenevinylene) rosettes and their tubular self-assembly, *Angew. Chem. Int. Ed.*, **43**, pp. 74–78.

15. Jin, S., Ma, Y., Zimmerman, S. C., and Cheng, S. Z. D. (2004). An ABC stacking supramolecular discotic columnar structure constructed via hydrogen-bonded hexamers, *Chem. Mater.*, **16**, pp. 2975–2977.
16. Yagai, S., Nakajima, T., Kishikawa, K., Kohmoto, S., Karatsu, T., and Kitamura, A. (2005). Hierarchical organization of photoresponsive hydrogen-bonded rosettes, *J. Am. Chem. Soc.*, **127**, pp. 11134–11139.
17. Yagai, S., Mahesh, S., Kikkawa, Y., Unoike, K., Karatsu, T., Kitamura, A., and Ajayaghosh, A. (2008). Toroidal nanoobjects from rosette assemblies of melamine-linked oligo(*p*-phenyleneethynylene)s and cyanurates, *Angew. Chem. Int. Ed.*, **47**, pp. 4691–4694.
18. Yagai, S., Usui, M., Seki, T., Murayama, H., Kikkawa, Y., Uemura, S., Karatsu, T., Kitamura, A., Asano, A., and Seki, S. (2012). Hierarchical organization of photoresponsive hydrogen-bonded rosettes, *J. Am. Chem. Soc.*, **134**, pp. 7983–7994.
19. Adhikari, B., Lin, X., Yamauchi, M., Ouchi, H., Aratsu, K., and Yagai, S. (2017). Hydrogen-bonded rosettes comprising π-conjugated systems as building blocks for functional one-dimensional assemblies, *Chem. Commun.*, **53**, pp. 9663–9683.
20. Yagai, S. (2015). Supramolecularly engineered functional π-assemblies based on complementary hydrogen-bonding interactions, *Bull. Chem. Soc. Jpn.*, **88**, pp. 28–58.
21. Yagai, S. (2006). Supramolecular complexes of functional chromophores based on multiple hydrogen-bonding interactions, *J. Photochem. Photochem. Photobiol. C*, **7**, pp. 164–182.
22. Seki, T., Lin, X., and Yagai, S. (2013). Supramolecular engineering of perylene bisimide assemblies based on complementary multiple hydrogen bonding interactions, *Asian J. Org. Chem.*, **2**, pp. 708–724.
23. Yagai, S., Kinoshita, T., Kikkawa, Y., Karatsu, T., Kitamura, A., Honsho, Y., and Seki, S. (2009). Interconvertible oligothiophene nanorods and nanotapes with high charge-carrier mobilities, *Chem. Eur. J.*, **15**, pp. 9320–9324.
24. Yagai, S., Goto, Y., Lin, X., Karatsu, T., Kitamura, A., Kuzuhara, D., Yamada, H., Kikkawa, Y., Saeki, A., and Seki, S. (2012). Self-organization of hydrogen-bonding naphthalene chromophores into J-type nanorings and H-type nanorods: Impact of regioisomerism, *Angew. Chem. Int. Ed.*, **51**, pp. 6643–6647.
25. Yagai, S., Kubota, S., Saito, H., Unoike, K., Karatsu, T., Kitamura, A., Ajayaghosh, A., Kanesato, M., and Kikkawa, Y. (2009). Reversible transformation between rings and coils in a dynamic hydrogen-bonded self-assembly, *J. Am. Chem. Soc.*, **131**, pp. 5408–5410.

26. Yagai, S., Goto, Y., Karatsu, T., Kitamura, A., and Kikkawa, Y. (2011). Catenation of self-assembled nanorings, *Chem. Eur. J.*, **17**, pp. 13657–13660.
27. Adhikari, B., Yamada, Y., Yamauchi, M., Wakita, K., Lin, X., Aratsu, K., Ohba, T., Karatsu, T., Hollambly, M. J., Shimizu, N., Takagi, H., Haruki, R., Adachi, S., and Yagai, S. (2017). Light-induced unfolding and refolding of supramolecular polymer nanofibers, *Nat. Commun.*, **8**, pp. 15254.
28. González-Rodríguez, D. and Schenning, A. P. H. J. (2011). Hydrogen-bonded supramolecular π-functional materials, *Chem. Mater.*, **23**, pp. 310–325.
29. Yagai, S., Suzuki, M., Lin, X., Gushiken, M., Noguchi, T., Karatsu, T., Kitamura, A., Saeki, A., Seki, S., Kikkawa, Y., Tani, Y., and Nakayama, K. (2014). Supramolecular engineering of oligothiophene nanorods without insulators: Hierarchical association of rosettes and photovoltaic properties, *Chem. Eur. J.*, **20**, pp. 16128–16137.
30. Lin, X., Suzuki, M., Gushiken, M., Yamauchi, M., Karatsu, T., Kizaki, T., Tani, Y., Nakayama, K., Suzuki, M., Yamada, H., Kajitani, T., Fukusihma, T., Kikkawa, Y., and Yagai, S. (2017). High-fidelity self-assembly pathways for hydrogen-bonding molecular semiconductors, *Sci. Rep.*, **7**, pp. 43098.
31. Miura, A., Jonkheijm, P., De Feyter, S., Schenning, A. P. H. J., Meijer, E. W., and De Schryver, F. C. (2004). 2D Self-assembly of oligo(*p*-phenylene vinylene) derivatives: From dimers to chiral rosettes, *Small*, **1**, pp. 131–137.
32. Ouchi, H., Lin, X., Kizaki, T., Prabhu, D. D., Silly, F., Kajitani, T., Fukushima, T., Nakayama, K., and Yagai, S. (2016). Hydrogen-bonded oligothiophene rosettes with a benzodithiophene terminal unit: Self-assembly and application to bulk heterojunction solar cells, *Chem. Commun.*, **52**, pp. 7874–7877.
33. Ouchi, H., Kizaki, T., Lin, X., Prabhu, D. D., Hoshi, N., Silly, F., Nakayama, K., and Yagai, S. (2017). Effect of alkyl substituents on 2D and 1D self-assembly and photovoltaic properties of hydrogen-bonded oligothiophene rosettes, *Chem. Lett.*, **46**, pp. 1102–1104.

Chapter 5

Fundamental Optoelectronic Process in Polymer:Fullerene Heterojunctions

Akinori Saeki

Department of Applied Chemistry, Graduate School of Engineering, Osaka University, 2-1 Yamadaoka, Suita, Osaka 565-0871, Japan
saeki@chem.eng.osaka-u.ac.jp

A power conversion efficiency (PCE) of organic photovoltaic (OPV) cell is the consequence of complex photophysical, morphological, and electric factors, which are hardly resolved into each component. This chapter accounts for separation, recombination, and transport of charge carriers in conjugated-polymer:fullerene OPVs revealed by using flash-photolysis time-resolved microwave conductivity (TRMC). This technique provides a direct access to inherent optoelectronics without electrodes, envisioning the wide prospect of TRMC in detailing charge-carrier dynamics. A fundamental of microwave spectroscopy is reviewed, and its advantages are exemplified by two recent studies on charge dynamics at the interface of orientation-controlled polymer and fullerene heterojunction and hole relaxation during transport in a bulk heterojunction film.

5.1 Introduction

A bulk heterojunction (BHJ) framework greatly benefits the efficient migration of exciton to the donor–acceptor interface and prevents non-geminate charge recombination (CR) in OPV devices [1]. Despite the excellent versatility of BHJs in polymer:fullerene [2], molecule:fullerene [3], polymer:polymer [4], and ternary blends [5], their photophysical processes intimately associated with the nanometer-scale morphology and self-assembling nature are hidden in device outputs. It is commonly recognized that the device parameters of short-circuit current density (J_{sc}), open-circuit voltage (V_{oc}), and fill factor (FF) are in a trade-off relationship, which limits the PCE defined by $J_{sc} \times V_{oc} \times FF/P_{in}$ (P_{in} is the area-normalized incident solar power). However, the present PCEs of solar cells, including crystalline silicon (~27%), OPV (~11%), and ever-growing perovskite (~22%), are still lower than the single-cell Shockley–Queisser (SQ) limit, which provides the theoretical maximum of PCE (~32%) [6, 7]. Accordingly, revealing the precise picture of loss mechanism and developing a method to circumvent the issue are important toward boosting a PCE to the SQ limit.

The migration of exciton and charge separation (CS) spontaneously occur on the sub-ps and ps timescales. This is followed by geminate CR (or formation of charge transfer complex), which is the initial loss pathway on sub-ns timescale. On the ns to μs regime, charge transport inside the active layer and charge collection at the buffer/electrode take place, where the non-geminate recombination—a carrier-density-dependent bimolecular recombination—appears to be the second loss process. The presence of trapping sites reduces the charge-carrier mobility and transport efficiency: In particular, a deep trap causes trap-assisted Shockley–Read–Hall (SRH) recombination, which results in a significant drop in J_{sc} and V_{oc}.

The time resolution of TRMC is typically a few to one hundred nanoseconds, which corresponds to the time domain of charge transport and non-geminate recombination. Under the alternating current (AC) electric field of microwave, charge carriers (hole and electron) undergo oscillation motion, which causes the absorption of microwave (Fig. 5.1). Consequently, the transient photoconductivity is obtained from the time evolution in the power of continuous, probing microwave at the trigger of the excitation light pulse. This chapter reviews the background underlying the electrode-less microwave spectroscopy to investigate charge-carrier dynamics. The next sections explain how the contrasting polymer orientation affects the CS and CR at the polymer/fullerene heterojunction and how the polymer crystallinity moderates the hole mobility during transport from the bottom to the top of the BHJ layer.

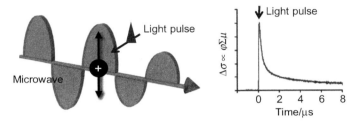

Figure 5.1 (Left) Illustration of interaction between microwave and charge carrier (hole) generated by a light pulse. (Right) Typical photoconductivity transient $\Delta\sigma$ (proportional to $\phi\Sigma\mu$).

5.2 Background of Microwave Spectroscopy

An electromagnetic wave (EMW), including radio wave, light, and X-ray radiation, is a form of energy that exhibits both wave-like and particle-like properties. The wave-like properties are identified by the wavelength (m) or frequency (Hz = s^{-1}), where their ranges are varied by orders of magnitudes. The ultraviolet, visible, and infrared lights are the most familiar EMWs that not only influence life on earth, but also have access to the electronic and structural properties of chemical species. The particle-like behavior of EMWs with 0.1–10 eV photon energy is essential for electronic excitation in molecular/atomic orbitals and subsequent photophysical and photochemical reactions. Infrared spectroscopy as well as Raman spectroscopy allows for detailing the chemical form of substituent and the weak hydrogen bond interaction, since the energy of infrared photon corresponds to the vibrational transition specific to the individual chemical bonds.

In contrast, the photon energy of gigahertz radiations (10^9 Hz = GHz) is of the order of 10^{-4} to 10^{-6} eV, even lower than the thermal energy at room temperature (0.025 eV). However, these energies match the rotational energy of polar molecules (e.g., H_2O), so that a microwave oven (e.g., 2.45 GHz microwave from a magnetron) and a microwave reactor are used to heat water, foods, and organic solvent without contacting a burning heat source. Microwave is a key component in wireless communication, radar, particle linear accelerator, and measurement (e.g., electron spin resonance, ESR).

Figure 5.2 shows the schematic of a TRMC setup composed of a light source and a microwave circuit, including source, waveguide, a resonant cavity, amplifier, and detector [8]. Owing to the moderate scale of wavelength (ca. 3 cm for 10 GHz X-band microwave), a resonant cavity is available to improve a signal-to-noise (S/N) ratio. The microwave frequency is tuned to a resonant frequency, which is determined by the cavity geometry and sample properties (size, position, and dielectric constant). The quality of the cavity (sensitivity and response time) is quantified by a Q value relating to the number of reflection. The microwave power from these sources is small and attenuated to a few to one hundred milliwatt, so that the electric field of the microwave neither disturbs the motion of charge carriers nor heats a sample. For the time-resolved measurement triggered by the excitation of light pulse, a wide range of light sources is available, such as an Nd:YAG laser (ca. 5–10 ns, repetition rate: 10 Hz) and visible light from an optical parametric oscillator seeded by the third harmonic generation of an Nd:YAG laser. The author has also utilized an Xe-flash lamp built in-house (10 µs pulse duration, 10 Hz) for a solar cell evaluation [9]. In addition, a high-energy electron beam pulse from an accelerator is another option (so-called pulse radiolysis), which allows "homogeneous" and "quantitative" generation of charges in non-polar solution [10] and block (powder) materials [11].

As a trade-off for the high sensitivity of the resonant cavity, the measurement frequency is limited to the specific frequency for each microwave circuit. Therefore, frequency dispersion is not obtained at one time as in terahertz time-domain spectroscopy. The author has developed frequency-modulated TRMC systems using K_u-band (ca. 15 GHz), K-band (ca. 23 GHz), and Q-band (or K_a-band, 33 GHz) in addition to the common X-band (ca. 9 GHz) [12]. The increase in resonant frequency (f_0) also merits the instrumental response time, as it is proportional to Q/f_0.

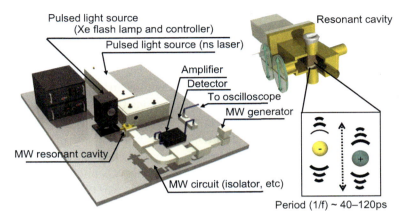

Figure 5.2 Schematic of a flash-photolysis TRMC system.

The real photoconductivity transient $\Delta\sigma$ is converted to the product of the quantum efficiency ϕ and the sum $\Sigma\mu$ ($= \mu_h + \mu_e$) of charge-carrier mobilities by

$$\phi\sum\mu = \frac{1}{e \cdot I_0 \cdot F_{\text{Light}}} \cdot \Delta\sigma \quad (5.1)$$

where e, I_0, and F_{Light} are the unit charge of a single electron, the excitation photon density of laser (photons cm^{-2}), and a correction factor (m^{-1}). F_{Light} was calculated in the same manner with F, but the depth profile of exciting laser is convoluted [8].

$$\Delta\sigma = \frac{1}{A}\frac{\Delta P_r}{P_r} \quad (5.2)$$

$$A = \frac{\mp Q\left(1/\sqrt{R_0} \pm 1\right)}{\pi f_0 \varepsilon_0 \varepsilon_r} \quad (5.3)$$

where A and $\Delta P_r/P_r$ are the sensitivity factor and the change in the fractional microwave power observed in the reflection from the cavity at the resonant frequency, respectively [13]. R_0 is the ratio of reflected (P_r) and incident (P_i) microwave power ($R_0 = P_r/P_i$), and ε_r is the real part of the relative dielectric constant inside the cavity.

Since A is experimentally determined, transient decays of $\Delta\sigma$ and $\phi\Sigma\mu$ are readily evaluated for each sample.

5.3 Charge Separation and Recombination at the Interface of Polymer/Fullerene Bilayer

A conjugated polymer is like a long belt with a surface and an edge side. The surface is furnished with π-electrons that are responsible for the optical and electronic response, while the edge side is usually hydrocarbons, which act as an insulator but are indispensable for solubilizing the rigid π-conjugated backbone. Therefore, two types of polymer:fullerene heterojunction are assumed in a BHJ film: face-on and edge-on. By using TRMC, we illustrate how the polymer orientation impacts the CS and CR processes at the bilayer interface composed of the donor polymer (PCPDTBT) and the acceptor fullerene (phenyl-C_{61}-butyric acid methyl ester, PCBM) (Fig. 5.3) [14]. To provide a model system of the identical backbone, the polymer orientation is controlled by the side alkyl chains of the cychlopentadithiophene (CPDT) unit, leading to face-on-rich for 2-ethylhexyl (EH) and edge-on for n-dodecyl (C_{12}) and n-hexadecanyl (C_{16}). PCPDTBT-EH is a milestone low band-gap polymer (LBP) of OPV showing PCEs of ~3% for the films processed from o-dichlorobenzene [15] and ~5% for the films processed from chlorobenzene (CB) with a high-boiling-point additive (1,8-diiodooctane: DIO) [16]. In contrast, PCPDTBT-C_{16} has demonstrated respectful field-effect transistor (FET) hole mobilities as large as 31 cm^2/Vs [17]. The fine control over face-on and edge-on orientations is a versatile approach toward efficient OPV and FET, respectively.

To simplify the model bilayer films, we prepared a PCPDTBT film (ca. 50 nm thick) on a PCBM layer using the contact film transfer (CFT) method [18]. The orientation of the pristine polymer film was confirmed by two-dimensional (wide-angle) grazing-incidence X-ray diffraction (2D-GIXRD), where EH shows face-on-rich orientation (mixture of face-on and edge-on) and C_{12} and C_{16} show mostly pure edge-on orientation.

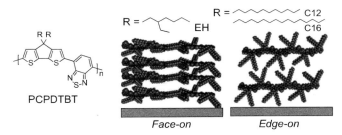

Figure 5.3 (Left) Chemical structure of PCPDTBT. (Center) Schematic of face-on orientation for PCPDTBT-EH. (Right) Schematic of edge-on orientation for PCPDTBT-C$_{12}$ and PCPDTBT-C$_{16}$.

Photophysical processes in the single layer of polymer and bilayer of polymer/PCBM were evaluated by flash-photolysis TRMC using a 650 nm nanosecond pulse as the excitation. The films were exposed from the polymer side, and thus the 650 nm light is absorbed mainly by the polymer. The transient TRMC signals ($\phi\Sigma\mu$) of single and bilayer films are shown in Fig. 5.4. Given that only hole mobility (μ_h) in pristine polymer film contributes to $\Sigma\mu$, we independently determined the μ_h values of EH, C$_{12}$, and C$_{16}$ to be 0.28 ± 0.03, 0.78 ± 0.07, and 1.2 ± 0.1 cm^2/Vs, respectively, by comparing with the transient photocurrent experiments. All the bilayer films showed increased photoconductivity maxima ($\phi\Sigma\mu_{max}$) compared to the polymer single layer films. Thus, the increase in $\phi\Sigma\mu_{max}$ ($\Delta\phi\Sigma\mu_{max}$) is attributed to the increase in ϕ (= $\Delta\phi$ > 0) at the DA heterojunction and contribution of electron mobility (μ_e) in the PCBM layer. The μ_e in the solid-state PCBM has been evaluated to be 0.1 cm^2/Vs [19]. Accordingly, $\Delta\phi$ is evaluated from the experimental $\Delta\phi\Sigma\mu_{max}$, μ_h, and μ_e as follows:

$$\Delta\phi = \frac{\Delta\phi\sum\mu_{max}}{\mu_h + \mu_e} \tag{5.4}$$

The $\Delta\phi$'s are as small as (6.8 ± 0.7) × 10^{-4} for EH/PCBM, (6.6 ± 0.7) × 10^{-4} for C$_{12}$/PCBM, and (2.4 ± 0.2) × 10^{-4} for C$_{16}$/PCBM. They result from the sequential processes, including exciton quenching, exciton diffusion, CS, and fast CR, all of which occur faster than the instrumental response time of TRMC. In particular, exciton diffusion yield is reduced in a DA bilayer system compared to BHJ.

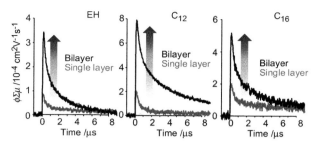

Figure 5.4 TRMC transients of single PCPDTBT films (gray) and PCPDTBT/PCBM bilayer films (black). (Left) EH, (center) C_{12}, and (right) C_{16} alkyl chains. The excitation wavelength was 650 nm. Reproduced with permission from Ref. [14], Copyright 2016, American Chemical Society.

A large portion of the singlet excitons is depleted under the high photon density of nanosecond laser via the first-order spontaneous decay [k_1 = 4.5 × 10^{-9} s^{-1} = (τ = 220 ps)$^{-1}$] [20] and high-order processes such as exciton–exciton [21] and exciton–polaron [22] quenching. The singlet exciton concentration $N(x, t)$ as a function of position x and time t is formulated by one-dimensional diffusion theory with the addition of the first- and second-order decay terms as follows [23]:

$$\frac{\partial N(x,t)}{\partial t} = D\frac{\partial^2 N(x,t)}{\partial x^2} - k_1 N(x,t) - k_2 N(x,t)^2 \qquad (5.5)$$

Whereas D is the diffusion constant reported for PCPDTBT (2.6 × 10^{-3} cm^2/s) and k_2 is the second-order rate constant of PCPDTBT (1.38 × 10^{-8} cm^3/s) [24], which is on the same order of typical rate constant of singlet-singlet annihilation in conjugated polymers, ~10^{-8} cm^3/s [25]. The initial and boundary conditions are given by

$$N(x,t)\big|_{t=0} = \begin{cases} \alpha I_0 10^{-\alpha x} \cdot \log 10 & (0 \le x \le L) \\ 0 & (x > L) \end{cases} \qquad (5.6)$$

$$D\frac{\partial N(x,t)}{\partial x}\bigg|_{x=0} = 0 \qquad (5.7)$$

where α is the absorption coefficient in the film states and I_0 is the laser photon density. To take into consideration the effect of pulse shape, the excitation nanosecond laser is assumed as a Gaussian (5 ns as full-width at half-maximum = 2.13 ns as the standard deviation

σ_L). By applying calculus of difference method to Eqs. (5.5)–(5.7), spatiotemporal decays of singlet excitons in EH/PCBM bilayer were obtained. Based on these numerical solutions, the exciton diffusion yield ϕ_{ED} was calculated.

The assumed loss process is illustrated in Fig. 5.5a. As a consequence, ϕ_{ED} values were calculated to be 1.6×10^{-2} for EH, 2.0×10^{-2} for C_{12}, and 3.5×10^{-2} for C_{16} (Fig. 5.5b). Since CS follows the exciton diffusion process, the calculated ϕ_{ED}, in turn, gives the CS yield of free carrier (ϕ_{CS}) to be $(4.1 \pm 0.4) \times 10^{-2}$ for EH, $(3.4 \pm 0.3) \times 10^{-2}$ for C_{12}, and $(0.69 \pm 0.07) \times 10^{-2}$ for C_{16}, by assuming $\Delta\phi = \phi_{ED} \times \phi_{CS}$. These ϕ_{CS} values are approximately one order smaller than the reported external quantum efficiencies (EQEs) of ~0.5 for PCPDTBT(EH):PC$_{71}$BM [26] and ~0.3 for PCPDTBT(C_{12}):PC$_{71}$BM [27] as well as the charge dissociation yields (~0.7) evaluated by femtosecond transient absorption spectroscopy (TAS) [20]. This is mostly due to the involvement of CR in ϕ_{CS}, because ϕ_{CS} is the free charge carrier generation yield at the instrumental response function of the present TRMC measurement (ca. 40 ns at 9 GHz). It is noteworthy that the ϕ_{CS} of face-on-rich EH (4.1×10^{-2}) is higher than that of edge-on C_{16} (0.69×10^{-2}), while the difference between EH and edge-on C_{12} (3.4×10^{-2}) is not obvious. This is indicative of the efficient CS arisen from the proximate polymer backbone to PCBM.

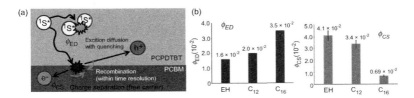

Figure 5.5 (a) Schematic of assumed process from exciton generation, exciton diffusion with quenching, and charge separation (free carrier) within the response time. (c) (Left) Calculated ϕ_{ED} and (left) ϕ_{CS} estimated from the experimental $\Delta\phi$ and calculated ϕ_{ED}. Reproduced with permission from Ref. [14], Copyright 2016, American Chemical Society.

Despite much lower ϕ_{CS} of C_{16}, the decay speed of TRMC transient is decelerated in comparison with EH. The results indicate that long-separated π-cores of donor and acceptor reduce bulk CR. This can enhance charge collection without current loss, while the CS yield is considerably decreased. Consequently, the overall performance of solar cell is superior for face-on compared to that of edge-on heterojunction. The correlation between the PCE/V_{oc} of OPV devices and the local charge-carrier mobilities ($\Sigma\mu$) of polymer:PCBM BHJ films has been previously reported [9], which is further assumed to relate with the crystallite sizes of polymer domain estimated by XRD [19]. It is concluded that the DA interface configuration, whether it is face-on or edge-on, rather than a high local mobility associated with charge delocalization, governs the CS yield. The local hole mobility of 0.28 cm^2/Vs of EH is much lower than those of C_{12} and C_{16} (~1 cm^2/Vs), while the ϕ_{CS} of the former is six-fold higher than that of C_{16}, suggestive of utmost importance of controlling DA interface and distance.

5.4 Charge Transport in Polymer:Fullerene Bulk Heterojunction

Charge-carrier mobility is a crucial parameter for electronic devices, which determines current flow and a balance of positive and negative charges; it is also associated with the CS at the donor–acceptor interface. Typical techniques to evaluate non-temporal charge-carrier mobility include space-charge-limited current (SCLC), FET, and time-of-flight (TOF) measurements, which are referred to as DC methods. In contrast, time-dependent mobilities (or conductivities) are evaluated by AC methods, e.g., time-resolved terahertz spectroscopy (TRTS) [28], TRMC using gigahertz microwaves [29, 30], and electric-field-induced second harmonic (EFISH) [31]. TRTS and time-resolved gigahertz spectroscopy constitute non-contact measurements, which can avoid contact issues; however, these methods are mostly incompatible with a device comprising metal or conducting electrodes. We have developed a new TOF–TRMC system, which allows measurement of hole relaxation in BHJ films. Transient photocurrent (TPC) is measured in a layered device under an external bias to evaluate conventional TOF mobilities [32]. TRMC transients are simultaneously measured in an identical setup.

Figure 5.6a shows the schematic of a TOF–TRMC experimental setup with an X-band harmonic resonant cavity. Before deciding on this cavity design, a layered device was examined in a reflection-type, rectangular TE$_{102}$-mode cavity (Q value ~2500 in an empty cavity), which has been typically used. However, resonant frequency was not observed because of the high reflectance and/or absorption of an ITO layer. The use of a transparent conductive oxide (TCO) is indispensable for efficient light-triggered charge injection and the in operando evaluation of TRMC transients under an external bias. The new, passing-type harmonic cavity (TE$_{10m}$, m = 14) affords Q values of ~3300 and ~450 in an empty cavity and with a device, respectively, which permits the simultaneous evaluation of TOF and TRMC. Figure 5.6b shows the device structure, where a BHJ layer (~2 μm thickness) is fabricated on ZnO/ITO/quartz, and an Au counter electrode is deposited by thermal evaporation.

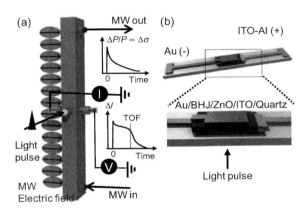

Figure 5.6 (a) Schematic of a system for the simultaneous measurements of TOF and TRMC using a harmonic cavity and a layered device. Blue ellipsoids represent the electric field generated by microwaves in the cavity. (b) Device structure for the TOF–TRMC measurement. Reproduced with permission from Ref. [32], Copyright 2017, American Chemical Society.

Figure 5.7a shows the TPC profiles (ΔI) under external electric field (E) of 10^3–10^4 V/cm. All curves exhibit typical dispersive decay with a kink point in the logarithmic plots, corresponding to τ_{TOF}. With increasing E, the intensity of ΔI increases, while τ_{TOF} decreases, both of which are directly related to the increase in the hole velocity at high voltage. In contrast, the intensity of the TRMC signal ($\Delta P/P$: change in microwave power divided by its baseline) mostly remains constant, while decay is accelerated with increasing E (Fig. 5.7b). The dependence on bias is more pronounced for crystalline polymers (i.e., P3HT and PffBT4T) compared to semi-crystalline or rather amorphous polymer (PCPDTBT). The latter exhibits a prompt decay (<1 μs), with no distinct bias dependence. Note that holes still survive after a delay of 1 μs as the TOF transients still exhibit a long tail at greater than 10 μs.

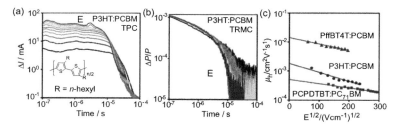

Figure 5.7 (a) Logarithmic plots of TPC profiles of P3HT:PCBM. The arrow corresponds to the increase in the electric field strength (E) from 5.6×10^3 to 5.6×10^4 V/cm. (b) Logarithmic plots of TRMC decays of P3HT:PCBM. The TPC and TRMC transients were simultaneously measured using the same laser at the same excitation intensity. (c) Hole mobilities (μ_h) as a function of the square root of electric field strength (E). Reproduced with permission from Ref. [32], Copyright 2017, American Chemical Society.

Figure 5.7c shows the E-dependent hole mobilities, μ_h [= $L(E \, \tau_{TOF})^{-1}$, where L is film thickness]. The extrapolated zero-field mobilities, μ_h ($E = 0$), are 1.8×10^{-3} cm^2/Vs for P3HT:PCBM, 1.4×10^{-2} cm^2/Vs for PffBT4T:PCBM, and 5.1×10^{-4} cm^2/Vs for PCPDTBT:PC$_{71}$BM. The μ_h ($E = 0$) of P3HT:PCBM is a few to tenfold greater than those evaluated by SCLC (2–6.4 × 10^{-4} cm^2/Vs) [33], but it is in agreement with previously reported TOF values [34]. PffBT4T:PCBM and PCPDTBT:PC$_{71}$BM exhibit μ_h values almost comparable with those reported previously by SCLC or FET measurements (3 × 10^{-2} cm^2/Vs for PffBT4T:PCBM [35] and 3 × 10^{-4} cm^2/Vs for PCPDTBT:PC$_{71}$BM) [36]. The negative dependence of μ_h on $E^{1/2}$ observed in all polymers suggests that holes are compelled to follow an unfavorable path with spatial disorder under external bias and not the thermally activated Poole–Frenkel framework (e.g., electrons in P3HT:PCBM) [37]. The negative dependence of holes has been reported in P3HT:PCBM [34].

TRMC decays are considerably different from those of TPC, although the charge carriers are subjected to the same bias in identical devices. The TPC profiles of P3HT and PffBT4T are relatively flat at shorter than τ_{TOF}, which progressively decrease beyond this time. In contrast, the TRMC curves converge to the baseline at around τ_{TOF}. This convergence is readily attributed to the collection of holes at the Au electrode. TRMC decays are always more rapid than TPC decays at less than τ_{TOF}, suggesting that the local hole mobilities probed by gigahertz electromagnetic waves undergo relaxation. Alternatively, non-geminate, bulk CR can also lead to the decay of the TRMC transient, but does not contribute to the TPC. To consider bulk recombination, two differential equations based on one-dimensional diffusion under an external electric field ($E > 0$) and the second-order recombination of holes and electrons, respectively, are formulated.

$$\frac{\partial n_p}{\partial t} = D_p \frac{\partial^2 n_p}{\partial x^2} - \gamma \cdot n_p n_n - E\mu_p \frac{\partial n_p}{\partial x} \tag{5.8}$$

$$\frac{\partial n_n}{\partial t} = D_n \frac{\partial^2 n_n}{\partial x^2} - \gamma \cdot n_p n_n + E\mu_n \frac{\partial n_n}{\partial x} \tag{5.9}$$

Here, n_p (n_n) is the density of holes (electrons), D_p (D_n) is the diffusion constant of holes (electrons), which is related to its mobility via Einstein's relation ($\mu_p = eD_p/k_BT$, where e is the unit charge of a single electron, k_B is the Boltzmann constant, T is the absolute temperature, 300 K), and γ is the bimolecular (second-order) rate constant. The zero-field hole mobility obtained by TOF experiments was used to calculate D_p. In the same manner, μ_p was varied according to the E-dependent TOF mobilities. The D_n and μ_n values of electrons in the polymer:PCBM blends are assumed to be constant; their values have been taken from previously reported studies: $\mu_n = 1 \times 10^{-3}$ cm^2/Vs for P3HT:PCBM [33], 8.1×10^{-4} cm^2/Vs for PffBT4T:PCBM [35], and 4×10^{-4} cm^2/Vs for PCPDTBT:PC$_{71}$BM [36]. Notably, electron mobility is not a decisive parameter because electrons are mostly stationary at the surface and are immediately collected by the ZnO/ITO electrode after pulse exposure. The initial condition regarding the hole and electron distributions is expressed by the Beer–Lambert law (decadic absorption coefficients ε are appended in the experimental section) as follows:

$$n_p(x, t = 0) = n_n(x, t = 0) = \phi_{EOP} \cdot \varepsilon I_0 10^{-\varepsilon x} \cdot \log 10 \tag{5.10}$$

where ϕ_{EOP} is the charge generation yield at the end of pulse (time resolution of the present TRMC). The boundary conditions constitute the extraction of electrons and the reflection of holes at $x = 0$ and n_p (n_n) = 0 at $x = \infty$ (the calculations were stopped before the holes reach the Au electrode). The γ values are taken from the studies that evaluate the identical polymer:$PC_{(71)}BM$ blend films at the low excitation by using transient absorption spectroscopy ($\gamma = 1.5 \times 10^{-12}$ cm^3/s for P3HT:PCBM [38], 1×10^{-11} cm^3/s for PffBT4T:PCBM [39], and 1×10^{-10} cm^3/s for PCPDTBT:$PC_{71}BM$ [40]). Accordingly, ϕ_{EOP} is a sole fitting parameter that is assumed to be constant regardless of the external bias.

Eqs. (5.8) and (5.9) were numerically solved by a finite difference method, and ϕ_{EOP} was screened to minimize the residual sum of squares for the TRMC decay at the lowest E. Figure 5.8a shows the fit of the hole decay in P3HT:PCBM at $E = 5.6 \times 10^3$ V/cm, as well as the curves calculated using the same ϕ_{EOP} (1 × 10^{-3}) with increasing E (PffBT4T:PCBM $\phi_{EOP} = 1 \times 10^{-3}$, PCPDTBT:$PC_{71}BM$ $\phi_{EOP} = 3 \times 10^{-4}$). These small ϕ_{EOP} values at high I_0 are consistent with the previous reports ($\phi_{EOP} < 10^{-2}$) [19, 29]. The ratio of $(\Delta P/P/I_0)_{max}$ at $I_0 = 10^{16}$ cm^{-2} per pulse (the intensity of the present experiments) and 10^{11} cm^{-2} per pulse (non-geminate recombination within the time resolution is assumed negligible) is calculated to be ~10^{-3} by extrapolating the observed dependence. In addition, the yields of collected charge carrier (ϕ_{TPC}) obtained by integrating TPC profiles (ΔI) are found to be 2.4×10^{-4} to 5.9×10^{-4} at the lowest E. These values are close to the obtained ϕ_{EOP} values, indicating the consistency of our experiments and analysis.

The normalized probability is expressed by $N_p(t)/N_p(0)$, where $N_p(t)$ is the density of hole (cm^{-2}) integrated along the depth (x), which is calculated as follows:

$$N_p(t) = \int_0^\infty n_p(x,t)dx \tag{5.11}$$

$N_n(t)$ is also defined in the same manner ($N_p(0) = N_n(0)$). The initial speeds of decay at less than ~100 ns are similar, while the time to reach saturation is gradually shortened with increasing E. Figure 5.8b shows the spatiotemporal evolution of the hole and electron densities in P3HT:PCBM at high E. The electrons are promptly quenched via bulk recombination and collection at the electrode without any significant change from the initial exponential function. In sharp contrast, holes are moved to a longer distance under the external bias, and their distribution is transformed from the initial exponential shape to a Gaussian one. Accordingly, the saturation of $N_p(t)/N_p(0)$ at a delayed time is rationalized by the spatial separation of the holes and electrons under E, which suppresses CR.

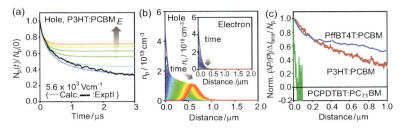

Figure 5.8 (a) Normalized probability of holes in P3HT:PCBM at $E = 5.6 \times 10^3$ V/cm (bottom) to $E = 5.6 \times 10^4$ V/cm (top). Black line represents the normalized TRMC decay under $E = 5.6 \times 10^3$ V/cm. (b) The reconstructed spatiotemporal profiles of hole and electron (inset) densities in P3HT:PCBM under $E = 5.6 \times 10^4$ V/cm. The color change with an arrow represents the elapsed time (blue: 0 s; red: 2.95 μs ≈ 1/3 τ_{TOF}). (c) The spatiotemporal profiles of normalized $(\Delta P/P)/\Delta I_{ana}/N_p$ under the highest E (red: 5.6 × 10^4 V/cm; blue: 3.9 × 10^4 V/cm; green: 8.7 × 10^4 V/cm). Reproduced with permission from Ref. [32], Copyright 2017, American Chemical Society.

The TRMC transients ($\Delta P/P$) are divided by the analyzed curve of TPC (ΔI_{ana}) and $N_p = N_p(t)/N_p(0)$ for compensating the decrease in holes as a result of the bulk recombination and trapping detected by TPC. The time is further normalized by τ_{TOF} to cancel the variations in film thickness, translational TOF mobility, and its E-dependence. The hole relaxation observed in the TOF–TRMC measurement is possibly related to the trapping and resultant lower shift in the mean mobility. Such a slow, but more pronounced, relaxation has been reportedly observed for PTB7:PC$_{71}$BM by integral-mode photocurrent measurements using a power law of relaxation ($\mu = \mu_0 t^{-\alpha}$) corresponding to trapping [41].

The E-dependence of $(\Delta P/P)/\Delta I_{ana}/N_p$ is in qualitative agreement with the negative dependence of TOF mobility, while the degree of the former is less significant than the latter. TOF–TRMC represents the simultaneous, independent evaluation of the translational and oscillation mobilities of charge carriers. Accordingly, $(\Delta P/P)/\Delta I_{ana}/N_p$

can be re-plotted against the travel distance by multiplying the time/τ_{TOF} and film thickness. Low-crystallinity PCPDTBT:PC$_{71}$BM exhibits a prompt decay down to the baseline at ~0.1 µm, while PffBT4T:PCBM and P3HT:PCBM exhibit moderate decay rates (Fig. 5.8c). Face-on-rich PffBT4T exhibits slower relaxation compared to edge-on P3HT. The film thickness is one of the most essential parameters for solar cell optimization. A thick film is preferred from the viewpoint of light absorption for increasing J_{sc}; however, FF and J_{sc} decrease for an extremely thick film because of the increased loss via charge trap and subsequent recombination. Notably, the degrees of relaxation observed in TOF–TRMC studies are in agreement with the optimal thickness of the solar cells: ~100 nm, ~200 nm, and ~300 nm for PCPDTBT [36], P3HT [42], and PffBT4T [43], respectively. Namely, the slow relaxation of holes permits maximization of optimal thickness. PffBT4T is a highly crystalline, face-on-rich polymer with a high hole mobility [43]. The exceptionally rapid decay observed for PCPDTBT:PC$_{71}$BM (~0 at 100 nm) is not solely explained by hole relaxation. The contribution of electron mobility to the TRMC signal is speculated to be responsible for the rapid decay of PCPDTBT:PC$_{71}$BM. However, the hole relaxation for PCPDTBT:PC$_{71}$BM is the most important reason for the complete quenching of TRMC signals observed at greater than 1 µs, although the TPC signals, solely related to the holes beyond this time, are still observed up to ~20 µs.

5.5 Summary

A fundamental of the TRMC technique has been reviewed, along with its application in the investigations on CS and CR in polymer/PCBM bilayer and charge transport in polymer:PCBM BHJ.

The face-on-rich and edge-on orientations of PCPDTBT are controlled by modifying the side chain of polymer, leading to higher ϕ_{CS} but larger decay rate of EH than those of linear alkyl ones. Accordingly, the overall PCE is suggested better for face-on-rich EH, due to the trade-off between CS and CR. The donor/acceptor interface plays a key role in CS and CR processes, which is more dominant than the local charge-carrier mobilities associated with the hole delocalization length in PCPDTBT polymers.

The hole relaxation dynamics in BHJ films (P3HT, PffBT4T, and PCPDTBT blended with PCBM) has been investigated by simultaneous measurements of TPC (ΔI) and TRMC ($\Delta P/P$) under an external electric field. TOF analyses have given hole mobilities of 10^{-4} to 10^{-2} cm^2/Vs. By dividing $\Delta P/P$ with TPC and hole decays calculated from the diffusion theory incorporating bulk CR under an external bias, the hole relaxation in the BHJ films as a function of normalized time (time/τ_{TOF}) and distance (0–1 µm) has been calculated. The results revealed that hole relaxation clearly follows the order of amorphous PCPDTBT, edge-on crystalline P3HT, and face-on-rich crystalline PffBT4T, which is in agreement with the optimal thickness of solar cell devices. It is anticipated that TRMC will provide the impetus upon which future research directions will be founded.

Acknowledgments

This work was supported by the Japan Society for the Promotion of Science (JSPS) KAKENHI: Grant-in-aid for Scientific Research (B) (Grant No. JP25288084), Grant-in-Aid for Challenging Exploratory Research (Grant No. JP15K13816), Grant-in-aid for Scientific Research (A) (Grant No. JP16H02285) and PRESTO program (Grant No. JPMJPR15N6) from JST of Japan. The author acknowledges Mr. Yoshiki Shimata at Osaka University for his great efforts to this work.

References

1. Yu, G., Gao, J., Hummelen, J. C., Wudl, F., and Heeger, A. J. (1995). Polymer photovoltaic cells: Enhanced efficiencies via a network of internal donor-acceptor heterojunctions, *Science*, **270**, pp. 1789–1791.
2. Xiao, S., Zhang, Q., and You, W. (2017). Molecular engineering of conjugated polymers for solar cells: An updated report, *Adv. Mater.*, **29**, pp. 1601391/1–8.
3. Huang, Y., Kramer, E. J., Heeger, A. J., and Bazan, G. C. (2014). Bulk heterojunction solar cells: Morphology and performance relationships, *Chem. Rev.*, **114**, pp. 7006–7043.
4. Chen, S., Liu, Y., Zhang, L., Chow, P. C. Y., Wang, Z., Zhang, G., Ma, W., and Yan, H. (2017). A wide-bandgap donor polymer for highly efficient

non-fullerene organic solar cells with a small voltage loss, *J. Am. Chem. Soc.*, **139**, pp. 6298–6301.

5. Lu, L., Kelly, M. A., You, W., and Yu, L. (2015). Status and prospects for ternary organic photovoltaics, *Nat. Photo.*, **9**, pp. 491–500.

6. Green, M. A., Hishikawa, Y., Warta, W., Dunlop, E. D., Levi, D. H., Hohl-Ebinger, J., and Ho-Baillie, A. W. Y. (2017). Solar cell efficiency tables (version 50), *Prog. Photovolt. Res. Appl.*, **25**, pp. 668–676.

7. Shockley, W. and Queisser, H. (1961). Detailed balance limit of efficiency of p-n junction solar cells, *J. App. Phys.*, **32**, pp. 510–519.

8. Saeki, A., Seki, S., Sunagawa, T., Ushida, K., and Tagawa, S. (2006). Charge-carrier dynamics in polythiophene films studied by in-situ measurement of flash-photolysis time-resolved microwave conductivity (FP-TRMC) and transient optical spectroscopy (TOS), *Phil. Mag.*, **86**, pp. 1261–1276.

9. Saeki, A., Yoshikawa, S., Tsuji, M., Koizumi, Y., Ide, M., Vijayakumar, C., and Seki, S. (2012). A versatile approach to organic photovoltaics evaluation using white light pulse and microwave conductivity, *J. Am. Chem. Soc.*, **134**, pp. 19035–19042.

10. Grozema, F. C. and Siebbeles, L. D. A. (2011). Charge mobilities in conjugated polymers measured by pulse radiolysis time-resolved microwave conductivity: From single chains to solids, *J. Phys. Chem. Lett.*, **2**, pp. 2951–2958.

11. Patwardhan, S., Sengupta, S., Siebbeles, L. D. A., Würthner, F., and Grozema, F. C. (2012). Efficient charge transport in semisynthetic zinc chlorin dye assemblies, *J. Am. Chem. Soc.*, **134**, pp. 16147–16150.

12. Saeki, A., Yasutani, Y., Oga, H., and Seki, S. (2014). Frequency-modulated gigahertz complex conductivity of TiO_2 nanoparticles: Interplay of free and shallowly trapped electrons, *J. Phys. Chem. C*, **118**, pp. 22561–22572.

13. Infelta, P. P., de Haas, M. P., and Warman, J. M. (1977). The study of the transient conductivity of pulse irradiated dielectric liquids on a nanosecond timescale using microwaves, *Radiat. Phys. Chem.*, **10**, pp. 353–365.

14. Shimata, Y., Ide, M., Tashiro, M., Katouda, M., Imamura, Y., and Saeki, A. (2016). Charge dynamics at heterojunction between face-on/edge-on PCPDTBT and PCBM bilayer: Interplay of donor/acceptor distance and local charge carrier mobility, *J. Phys. Chem. C*, **120**, pp. 17887–17897.

15. Mühlbacher, D., Scharber, M., Morana, M., Zhu, Z., Waller, D., Gaudiana, R., and Brabec, C. (2006). High photovoltaic performance of a low-bandgap polymer, *Adv. Mater.*, **18**, pp. 2884–2889.

16. Lee, J. K., Ma, W. L., Brabec, C. J., Yuen, J., Moon, J. S., Kim, J. Y., Lee, K., Bazan, G. C., and Heeger, A. J. (2008). Processing additives for improved efficiency from bulk heterojunction solar cells, *J. Am. Chem. Soc.*, **130**, pp. 3619–3623.

17. Luo, C., Kyaw, A. K. K., Perez, L. A., Patel, S., Wang, M., Grimm, B., Bazan, G. C., Kramer, E. J., and Heeger, A. J. (2014). General strategy for self-assembly of highly oriented nanocrystalline semiconducting polymers with high mobility, *Nano Lett.*, **14**, pp. 2764–2771.

18. Wei, W., Miyanishi, S., Tajima, K., and Hashimoto, K. (2009). Enhanced charge transport in polymer thin-film transistors prepared by contact film transfer method, *ACS Appl. Mater. Interfaces*, **1**, pp. 2660–2666.

19. Yoshikawa, S., Saeki, A., Saito, M., Osaka, I., and Seki, S. (2015). On the role of local charge carrier mobility in the charge separation mechanism of organic photovoltaics, *Phys. Chem. Chem. Phys.*, **17**, pp. 17778–17784.

20. Yamamoto, S., Ohkita, H., Benten, H., and Ito, S. (2012). Role of interfacial charge transfer state in charge generation and recombination in low-bandgap polymer solar cell, *J. Phys. Chem. C*, **116**, pp. 14804–14810.

21. Dicker, G., de Haas, M. P., and Siebbeles, L. D. A. (2005). Signature of exciton annihilation in the photoconductance of regioregular poly(3-hexylthiophene), *Phys. Rev. B*, **71**, pp. 155204/1–8.

22. Hodgkiss, J. M., Albert-Seifried, S., Rao, A., Barker, A. J., Campbell, A. R., Marsh, R. A., and Friend, R. H. (2012). Exciton-charge annihilation in organic semiconductor films, *Adv. Funct. Mater.*, **22**, pp. 1567–1577.

23. Mikhnenko, O. V., Ruiter, R., Blom, P. W. M., and Loi, M. A. (2012). Direct measurement of the triplet exciton diffusion length in organic semiconductors, *Phys. Rev. Lett.*, **108**, pp. 137401/1–5.

24. Mikhnenko, O. V., Azimi, H., Scharber, M., Morana, M., Blom, P. W. M., and Loi, M. A. (2012). Exciton diffusion length in narrow bandgap polymers, *Energy Environ. Sci.*, **5**, pp. 6960–6965.

25. Cook, S., Liyuan, H., Furube, A., and Katoh, R. (2010). Singlet annihilation in films of regioregular poly(3-hexylthiophene): Estimates for singlet diffusion lengths and the correlation between singlet annihilation rates and spectral relaxation, *J. Phys. Chem. C*, **114**, pp. 10962–10968.

26. Peet, J., Kim, J. Y., Coates, N. E., Ma, W. L., Moses, D., Heeger, A. J., and Bazan, G. C. (2007). Efficiency enhancement in low-bandgap polymer solar cells by processing with alkane dithiols, *Nat. Mater.*, **6**, pp. 497–500.

27. Coffin, R. C., Peet, J., Rogers, J., and Bazan, G. C. (2009). Streamlined microwave-assisted preparation of narrow-bandgap conjugated

polymers for high performance bulk heterojunction solar cells, *Nat. Chem.*, **1**, pp. 657–661.

28. Jin, Z., Gehrig, D., Dyer-Smith, C., Heilweil, E. J., Laquai, F., Bonn, M., and Turchinovich, D. (2014). Ultrafast terahertz photoconductivity of photovoltaic polymer–fullerene blends: A comparative study correlated with photovoltaic device performance, *J. Phys. Chem. Lett.*, **5**, pp. 3662–3668.

29. Savenije, T. J., Ferguson, A. J., Kopidakis, N., and Rumbles, G. (2013). Revealing the dynamics of charge carriers in polymer:fullerene blends using photoinduced time-resolved microwave conductivity, *J. Phys. Chem. C*, **117**, pp. 24085–24103.

30. Saeki, A., Koizumi, Y., Aida, T., and Seki, S. (2012). Comprehensive approach to intrinsic charge carrier mobility in conjugated organic molecules, macromolecules, and supramolecular architectures, *Acc. Chem. Res.*, **45**, pp. 1193–1202.

31. Vithanage, D. A., Devižis, A., Abramavičius, V., Infahsaeng, Y., Abramavičius, D., MacKenzie, R. C. I., Keivanidis, P. E., Yartsev, A., Hertel, D., Nelson, J., Sundström, V., and Gulbinas, V. (2014). Visualizing charge separation in bulk heterojunction organic solar cells, *Nat. Commun.*, **4**, pp. 2334/1–6.

32. Shimata, Y. and Saeki, A. (2017). Hole relaxation in polymer:fullerene solar cells examined by the simultaneous measurement of time-of-flight and time-resolved microwave conductivity, *J. Phys. Chem. C*, **121**, pp. 18351–18359.

33. Mihailetchi, V. D., Xie, H., de Boer, B., Koster, L. J. A., and Blom, P. W. M. (2006). Charge transport and photocurrent generation in poly(3-hexylthiophene):methanofullerene bulk-heterojunction solar cells, *Adv. Funct. Mater.*, **16**, pp. 699–708.

34. Kim, Y., Cook, S., Tuladhar, S. M., Choulis, S. A., Nelson, J., Durrant, J. R., Bradley, D. D. C., Giles, M., Mcculoch, I., Ha, C.-S., and Ree, M. (2006). A strong regioregularity effect in self-organizing conjugated polymer films and high-efficiency polythiophene:fullerene solar cells, *Nat. Mater.*, **5**, pp. 197–203.

35. Kumano, M., Ide, M., Seiki, N., Shoji, Y., Fukushima, T., and Saeki, A. (2016). A ternary blend of a polymer, fullerene, and insulating self-assembling triptycene molecules for organic photovolatics, *J. Mater. Chem. A*, **4**, pp. 18490–18498.

36. Scharber, M. C., Koppe, M., Gao, J., Cordella, F., Loi, M. A., Denk, P., Morana, M., Egelhaaf, H.-J., Forberich, K., Dennler, G., Gaudiana, R., Waller, D., Zhu, Z., Shi, X., and Brabec, C. J. (2010). Influence of the bridging atom

on the performance of a low-bandgap bulk heterojunction solar cell, *Adv. Mater.*, **22**, pp. 367–370.

37. Morfa, A. J., Nardes, A. M., Shaheen, S. E., Kopidakis, N., and van de Lagemaat, J. (2011). Time-of-flight studies of electron-collection kinetics in polymer:fullerene bulk-heterojunction solar cells, *Adv. Funct. Mater.*, **21**, pp. 2580–2586.

38. Etzold, F., Howard, I. A., Mauer, R., Meister, M., Kim, T.-D., Lee, K.-S., Baek, N. S., and Laquai, F. (2011). Ultrafast exciton dissociation followed by nongeminate charge recombination in PCDTBT:PCBM photovoltaic blends, *J. Am. Chem. Soc.*, **133**, pp. 9469–9479.

39. Baran, D., Kirchartz, T., Wheeler, S., Dimitrov, S., Abdelsamie, M., Gorman, J., Ashraf, R. S., Holliday, S., Wadsworth, A., Gasparini, N., Kaienburg, P., Yan, H., Amassian, A., Brabec, C. J., Durrant, J. R., and McCulloch, I. (2016). Reduced voltage losses yield 10% efficient fullerene free organic solar cells with >1 V open circuit voltages, *Energy Environ. Sci.*, **9**, pp. 3783–3793.

40. Etzold, F., Howard, I. A., Forler, N., Cho, D. M., Meister, M., Mangold, H., Shu, J., Hansen, M. R., Müllen, K., and Laquai, F. (2012). The effect of solvent additives on morphology and excited-state dynamics in PCPDTBT:PCBM photovoltaic blends, *J. Am. Chem. Soc.*, **134**, pp. 10569–10583.

41. Pranculis, V., Ruseckas, A., Vithanage, D. A., Hedley, G. J., Samuel, I. D. W., and Gulbinas, V. (2016). Influence of blend ratio and processing additive on free carrier yield and mobility in PTB7:PC71BM photovoltaic solar cells, *J. Phys. Chem. C*, **120**, pp. 9588–9594.

42. Mikie, T., Saeki, A., Masuda, H., Ikuma, N., Kokubo, K., and Seki, S. (2015). New efficient (thio)acetalized fullerene monoadducts for organic solar cells: Characterization based on solubility, mobility balance, and dark current, *J. Mater. Chem. A*, **3**, pp. 1152–1157.

43. Liu, Y., Zhao, J., Li, Z., Mu, C., Ma, W., Hu, H., Jiang, K., Lin, H., Ade, H., and Yan, H. (2014). Aggregation and morphology control enables multiple cases of high-efficiency polymer solar cells, *Nat. Commun.*, **5**, pp. 5293/1–8.

Chapter 6

Near-Infrared Dyes Based on Ring-Expanded Porphyrins with No *Meso*-Bridges

Tetsuo Okujima
Graduate School of Science and Engineering, Ehime University,
2-5 Bunkyo-cho Matsuyama, 790-8577, Japan
okujima.tetsuo.mu@ehime-u.ac.jp

6.1 Introduction

The chemistry of ring-expanded porphyrins has gained growing focus because of their properties potentially suitable for a number of novel practical applications [1–11]. Expanded, isomeric, and contracted porphyrins are named by a combination of the number of conjugated π-electrons, pyrroles, and *meso*-atoms. For example, [22] pentaphyrin(1.1.1.1.0) is commonly called sapphyrin (Fig. 6.1). Since 2000, Osuka et al. have reported a series of ring-expanded porphyrins from pentaphyrin to dodecaphyrin wherein the neighboring pyrroles are connected with *meso*-carbons [5, 6, 10]. These porphyrins show

Light-Active Functional Organic Materials
Edited by Hiroko Yamada and Shiki Yagai
Copyright © 2019 Jenny Stanford Publishing Pte. Ltd.
ISBN 978-981-4800-15-0 (Hardcover), 978-0-429-44853-9 (eBook)
www.jennystanford.com

the intense Soret bands at 500–900 nm and the weak Q bands at 700–1300 nm. Interestingly, [28]hexaphyrin(1.1.1.1.1.1) adopts single-sided twisted ring conformation with Möbius aromaticity, while [26]hexaphyrin(1.1.1.1.1.1) adopts double-sided rectangular ring conformation with Hückel aromaticity [12, 13].

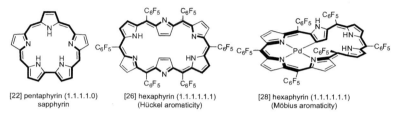

[22] pentaphyrin (1.1.1.1.0) sapphyrin   [26] hexaphyrin (1.1.1.1.1.1) (Hückel aromaticity)   [28] hexaphyrin (1.1.1.1.1.1) (Möbius aromaticity)

Figure 6.1 Structures of ring-expanded porphyrins.

In 2002, Sessler reported the first synthesis of cyclo[8]pyrrole, [30]octaphyrin(0.0.0.0.0.0.0.0), a ring-expanded porphyrin analogue with no *meso*-bridges, which showed porphyrinoid aromaticity contributed by a 30π-electron system as a highlighted bold line in Scheme 6.1 [14]. The structural difference from porphyrins is the complete absence of *meso*-carbons. The FeCl$_3$-induced oxidative coupling of 2,2′-bipyrrole in the presence of H$_2$SO$_4$ affords the corresponding cyclo[8]pyrroles **2** as a sulfate. The yields of **2a–c** exceeded 70%. The yield of **2d** became low due to the lack of full β-substitution of pyrrole moieties. The crystal structure of **2b** adopts a flat conformation with eight hydrogen-bonding interactions between pyrrolic NHs and the inner sulfate dianion. The absorption spectra of them show a weak B band at ca. 430 nm ($\varepsilon \approx 1 \times 10^5$ M^{-1} cm^{-1}) and a relatively intense L band at ca. 1100 nm ($\varepsilon \approx 2 \times 10^5$ M^{-1} cm^{-1}), which correspond to the Soret and Q bands of porphyrins. The strong absorbance in the near-infrared (NIR) region makes them potentially suitable for use in optical storage and signaling devices. The photophysical [6], anion-binding [15, 16], liquid-crystal properties [17], single-molecule transistor [18], and electronic structures [19, 20] of cyclo[8]pyrroles have been studied in depth. This section focuses on the synthesis of new derivatives of cyclo[8] pyrroles.

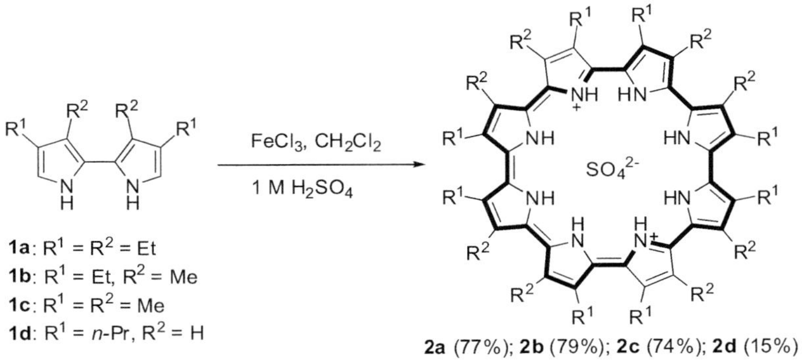

1a: R¹ = R² = Et
1b: R¹ = Et, R² = Me
1c: R¹ = R² = Me
1d: R¹ = n-Pr, R² = H

2a (77%); **2b** (79%); **2c** (74%); **2d** (15%)

Scheme 6.1 Synthesis of β-alkylcyclo[8]pyrroles.

6.2 π-Expanded Cyclo[8]pyrroles

The ability to fine-tune the wavelength of absorptions based on structural modification such as ring annelation would enhance the utility of these compounds for practical applications. Although various ring-annelated porphyrins have been synthesized [21, 22], there have been no reports of π-expanded cyclo[8]pyrroles previously. In 2011, Okujima, Kobayashi et al. successfully synthesized cyclo[8]isoindole, the first π-expanded cyclo[8]pyrrole, via the oxidative cyclization of bicyclo[2.2.2]octadiene (BCOD)-fused 2,2′-bipyrrole **3**, followed by thermal conversion (Scheme 6.2) [23]. The synthesis of **4** under different conditions is summarized in Table 6.1. The best yield was obtained by the use of Ce(SO$_4$)$_2$ as an oxidant, 6 M H$_2$SO$_4$ as an acid, and Na$_2$SO$_4$ and (n-Bu)$_4$NHSO$_4$ as additives because of the template effect of the SO$_4^{2-}$ anion. After purification by GPC and recrystallization, pure **4** was obtained as deep blue crystals. BCOD moieties are converted into fused-benzene rings by retro Diels–Alder reaction. When **4** was heated as a solid at 240°C under reduced pressure, the color changed from blue to yellow and **5** was formed in almost quantitative yield. Interestingly, the color of the solutions is different from the crystals (Fig. 6.2). The crystals of **4** and **5** are ground to same color as the solution, respectively. These observations indicate that **4** and **5** show mechanochromic behavior. Peripheral benzo-fusion results in a marked intensification and red-shift of the B band from 471 to 627 nm because of the stabilization of the LUMO+1. On the other hand, the L band of **5** is observed at 1078 nm similar to **4**.

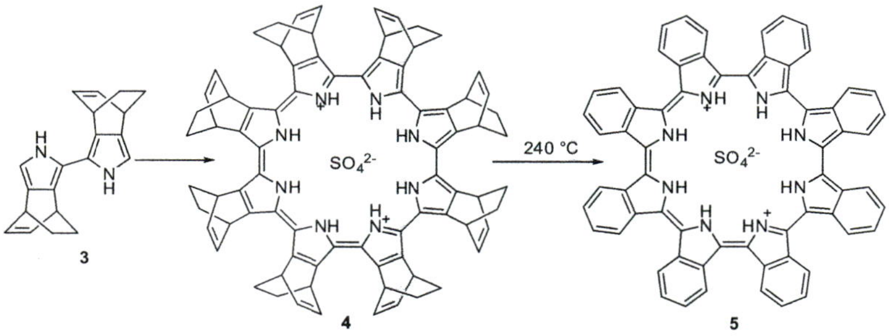

Scheme 6.2 Synthesis of cyclo[8]isoindole **5**.

Table 6.1 Oxidative coupling of **3**

Entry	Oxidant	Acid	Additive	Yield
1.	FeCl$_3$·6H$_2$O	1 M H$_2$SO$_4$	None	43%
2.	NaNO$_2$	Conc. H$_2$SO$_4$	None	40%
3.	CAN	AcOH	Na$_2$SO$_4$	28%
4.	AgO	Conc. HNO$_3$	Na$_2$SO$_4$	41%
5.	Ce(SO$_4$)$_2$	6 M H$_2$SO$_4$	Na$_2$SO$_4$	54%
6.	Ce(SO$_4$)$_2$	6 M H$_2$SO$_4$	Na$_2$SO$_4$, (n-Bu)$_4$NHSO$_4$	68%

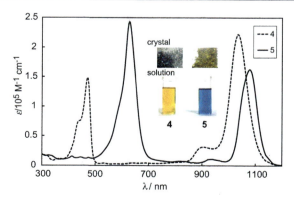

Figure 6.2 UV-vis-NIR absorption spectra of **4** (dotted line) and **5** (solid line). Pictures of single crystals and solutions are inserted.

Ring annelation of cyclo[8]pyrroles is based on two types of fused-ring expansions—**A**: 8 pyrroles fused with aromatic ring and **B**: 4 bipyrroles fused with aromatic ring, as shown in Fig. 6.3. Sessler and Panda subsequently reported synthesis of cyclo[4]naphthobipyrrole **7** (Scheme 6.3) [24, 25]. Cyclo[4]naphthobipyrroles (**7**) have red-shifted NIR L bands, which lie beyond λ = 1200 nm. The X-ray structures of **4**, **5**, and **7b** are shown in Fig. 6.4. The crystal structure of **4** is similar to that of **2** [14, 19]. Fused-ring expansion at the β-pyrrolic positions results in nonplanarity, a deeper saddling distortion of the π-system, due to the steric hindrance between the neighboring fused-benzene moieties of **5**. Cyclo[4]naphthobipyrrole **7** is free from this distortion to adopt a more planar conformation compared to **5**, which is why **7** shows the red-shifted L band.

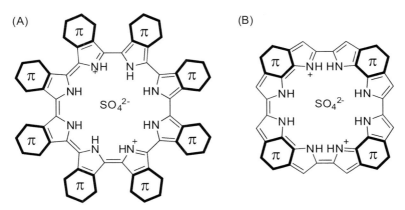

Figure 6.3 Structural models of π-expanded cyclo[8]pyrrole.

In 2013, acenaphthylene-fused cyclo[8]pyrroles were reported by Okujima, Kobayashi et al. (Scheme 6.4 and Table 6.2) [26]. Two conformational isomers were isolated by column chromatography on silica gel with CHCl$_3$ with R_f values of 0.6 for **9a** and 0.9 for **9b**. The yields of **9** depend on reaction time and temperature (entries 2–4), which suggests that less polar **9b** could be converted into **9a**. The molecular structures were elucidated by X-ray crystallographic analysis, as shown in Fig. 6.5. A solution of **9b** in o-dichlorobenzene-d_4 was heated at 144°C to afford **9a** through a thermal ring flip, monitored by NMR with signal broadening and coalescence. The obtained **9a** persists even after cooling to room temperature. The intense and marked red-shifted L bands of **9a** and **9b** are observed at 1489 nm (log ε 5.54) for **9a** and 1515 nm (log ε 5.34) for **9a** (Fig. 6.6).

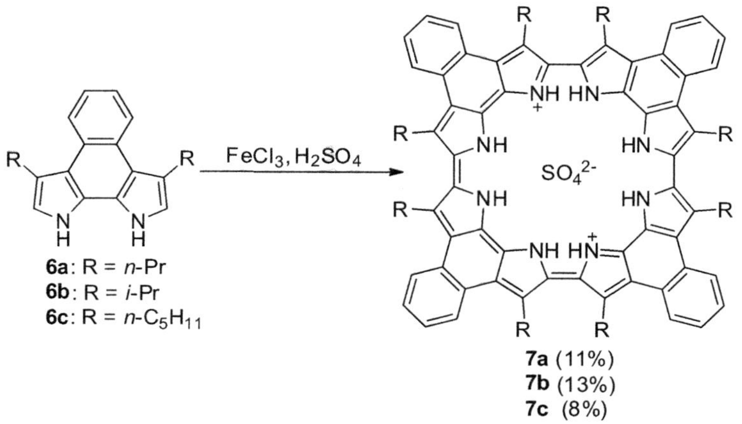

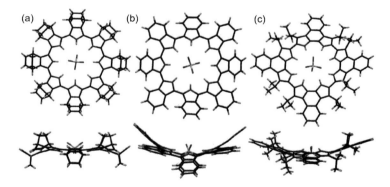

Scheme 6.3 Oxidative coupling of naphthobipyrrole.

Figure 6.4 X-ray structures of (a) **4**, (b) **5**, and (c) **7b**.

Table 6.2 Oxidative coupling of **8**

Entry	Oxidant	Acid	Additive	9a	9b
1[a]	FeCl$_3$·6H$_2$O	1 M H$_2$SO$_4$	None	7%	3%
2[b]	Ce(SO$_4$)$_2$	Conc. H$_2$SO$_4$	Na$_2$SO$_4$, (n-Bu)$_4$NHSO$_4$	20%	22%
3[c]	Ce(SO$_4$)$_2$	Conc. H$_2$SO$_4$	Na$_2$SO$_4$, (n-Bu)$_4$NHSO$_4$	38%	17%
4[d]	Ce(SO$_4$)$_2$	Conc. H$_2$SO$_4$	Na$_2$SO$_4$, (n-Bu)$_4$NHSO$_4$	43%	14%

[a]rt, 14 h; [b]rt, 4 h; [c]reflux, 4 h; [d]reflux, 27 h

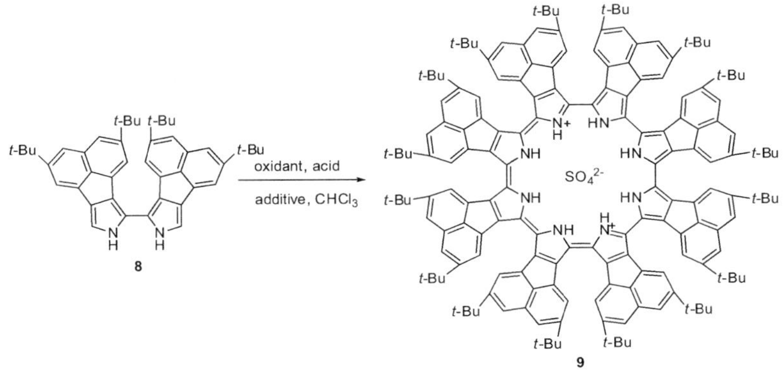

π-Expanded Cyclo[8]pyrroles | 121

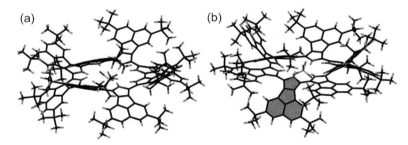

Scheme 6.4 Oxidative coupling of **8**.

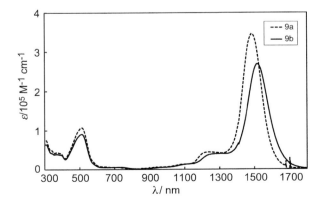

Figure 6.5 X-ray structures of (a) **9a** and (b) **9b**. The gray-colored acenaphthopyrrole moiety of **9b** flips to form alternative tilting structure of **9a**.

Figure 6.6 UV-vis-NIR absorption spectra of **9a** (dotted line) and **9b** (solid line).

Cyclo[8]pyrrole with eight 9,10-dihydroanthracene units (**11**) was synthesized by oxidative coupling of **10** [27]. Three-dimensional cyclo[8]pyrrole **11** possesses a cavity with a diameter of ca. 15 Å above and below porphyrin plane. A pair of stacked cyclo[8]pyrroles is observed from the X-ray crystal analysis.

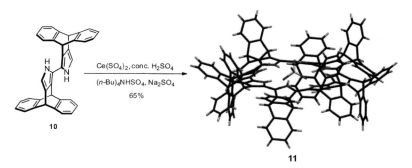

Scheme 6.5 Oxidative coupling of **10** and X-ray structure of **11**.

6.3 Dichloride, Carboxylate, and Polyoxometalate Salts

Cyclo[8]pyrroles are usually isolated as a sulfate salt of diprotonated octaphyrin. Little is known about the other salts of cyclo[8]pyrrole, despite the fact that porphyrin can coordinate with various metal ions into its central cavity. Dichloride salt can be isolated from oxidative coupling reaction in the presence of HCl [28]. In the case of reaction with H_2SO_4, cyclo[8]pyrrole is selectively obtained due to the template effect of the sulfate anion. Sessler et al. reported **12** along with smaller analogues, cyclo[7]pyrrole **13** (5%) and cyclo[6] pyrrole **14** (15%) (Scheme 6.6). In 2008, a supramolecular system of cyclo[8]pyrrole and pyrene carboxylate by anion-binding from **12** in the solution was reported as a photoinduced charge separation [16].

In 2016, Okujima et al. reported cyclo[8]pyrrole–polyoxometalate (POM) double-decker complex **15**, which was obtained by anion-exchange from **2a** (Scheme 6.7) [29]. The X-ray crystal analysis revealed the sandwich structure the POM tetra-anion held between two cyclo[8]pyrroles with weak hydrogen-bonding interaction. This double-decker complex is expected to be used as photofunctional materials based on a combination of POMs with porphyrinoids [30–32].

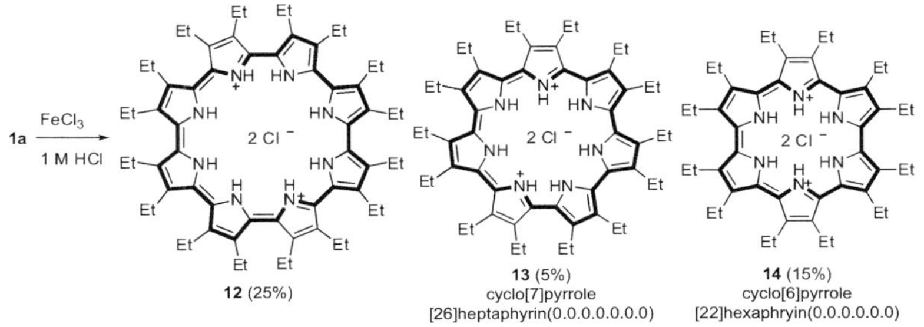

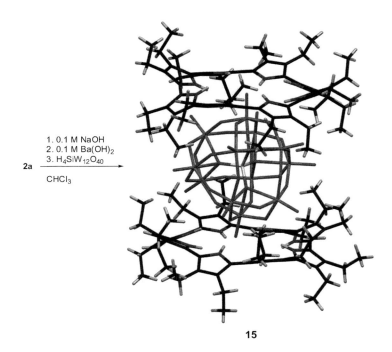

Scheme 6.6 Oxidative coupling of **1a**.

Scheme 6.7 Synthesis of double-decker complex **15**.

6.4 Smaller Analogues

Only two smaller analogues, cyclo[7]pyrrole **13** and cyclo[6]pyrrole **14**, were reported by Sessler et al. in 2003 [28]. They were isolated via column chromatography from the oxidative coupling of **1a** with FeCl$_3$ and HCl. The longest wavelength absorptions are observed at 936 nm for **13** and 792 nm for **14**.

Core-modified smaller analogues, cyclo[3]pyridine[3]pyrrole **16** and cyclo[2]pyridine[4]pyrrole **17**, were synthesized via Suzuki–Miyaura coupling (Fig. 6.7) [33]. Nonaromatic macrocycles **16** and **17** show localized aromaticity for pyridine and pyrrole moieties, whereas cyclo[6]pyrrole **14** contains a 22π-electron periphery of Hückel aromatic system. Diprotonation of **17** occurs readily to produce an extension of the π-conjugation. The diprotonated **17** is contributed by 24π-electron antiaromatic 2H**17**'$^{2+}$ as observed in the absorption spectrum and DFT calculations.

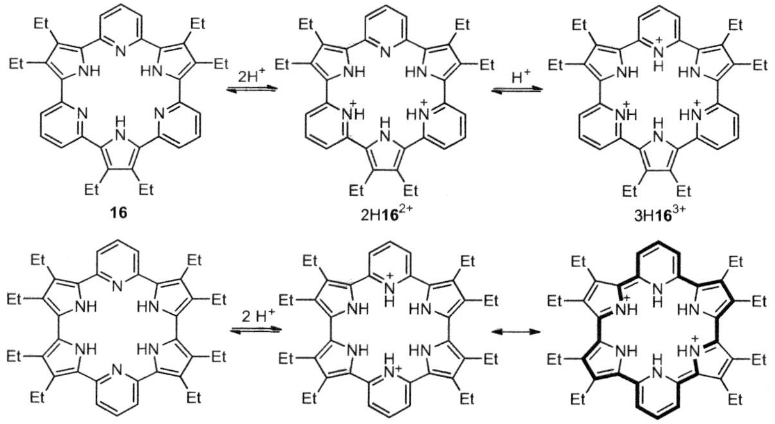

Figure 6.7 Structures of **16**, **17**, and their protonated species.

Cyclo[1]furan[1]pyridine[4]pyrrole **20** reported by Ishida, Kim, and Sessler is also prepared via Suzuki–Miyaura coupling of **18** and **19** (Scheme 6.8) [34]. Uranium (IV) treatment of **20** results in the formation of the anionic and oxidized form of **20**, [**20**–4H]$^{2-}$. In contrast to **16** and **17**, **21** possesses 22π-electron aromatic character as supported by spectroscopic, electrochemical, and theoretical studies.

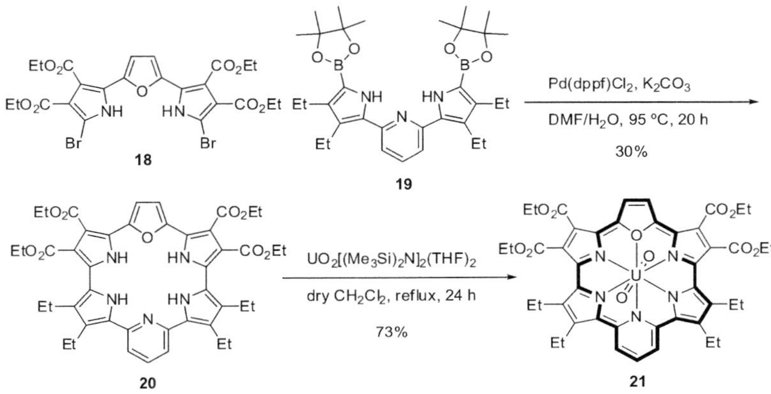

Scheme 6.8 Synthesis of cyclo[1]furan[1]pyridine[4]pyrroles **20** and **21**.

6.5 Larger Analogues

A core-modified larger analogue, thiophene-containing cyclo[9]pyrrole **22**, was obtained via mild electrochemical oxidation from 2,5-dipyrrolylthiophene (Fig. 6.8) [35]. Standard chemical oxidative coupling could not afford the corresponding cyclo[9]pyrrole. The π-system of **22** is assigned as a 34π-electron aromatic system on the basis of ^{1}H NMR spectroscopy. Sessler, Kim, and Ishida reported the larger cyclo[*n*]pyrrole analogue, cyclo[6]pyridine[6]pyrrole **23** synthesized via Suzuki–Miyaura cross-coupling in 2014 [36]. In its neutral form, **23** exists as a mixture of two twisted figure-eight conformers.

On the other hand, the mass spectra indicate the formation of cyclo[5]naphthobipyrrole **24** and cyclo[8]naphthobipyrrole in the polar fractions from column chromatography in the case of synthesis of cyclo[4]naphthobipyrrole **7b** [24]. However, these larger analogues were not isolated because of trace amounts of larger structural analogues formed under similar conditions.

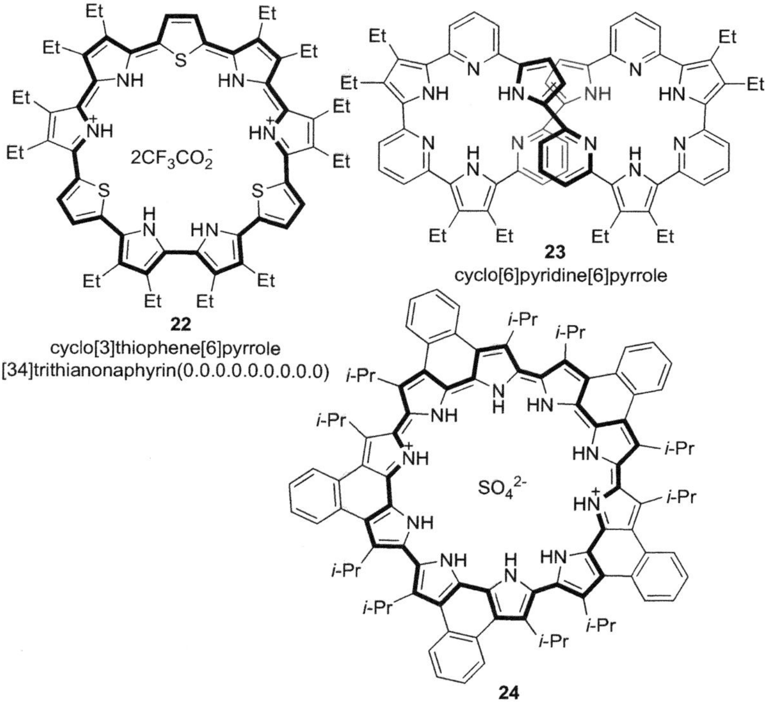

Figure 6.8 Structures of larger cyclo[*n*]pyrroles **22**, **23**, and **24**.

The formation of trace amounts of larger structural analogues was also indicated via the MALDI-TOF MS in the case of synthesis of cyclo[8]acenaphthopyrrole **9**. Although a cyclo[10]pyrrole skeleton was confirmed by X-ray crystallographic analysis after purification and isolation from the collected several fractions in different batches, it was not possible to determine the identity of dianion bound in the inner cavity. A selective synthesis of cyclo[8]pyrroles was achieved because of the template effect of the sulfate dianion. Thus, the size of the dianion template used is important to obtain the desired cyclo[*n*]pyrrole. Oxidative coupling of **8** in the presence of croconic acid, which is template dianion and a proton source, affords cyclo[10]pyrrole **25** selectively (Scheme 6.9) [37]. The largest macrocyclic cyclo[*n*]pyrrole shows the marked red-shifted L band at ca. 2000 nm (Fig. 6.9). While the weak B band is observed at 428 nm, the strong CT band from the core cyclo[10]pyrrole to the peripheral anenaphthylene moiety is observed between the B and L bands.

25 (66%) red crystals
cyclo[10]pyrrole
[38]decaphyrin(0.0.0.0.0.0.0.0.0.0)

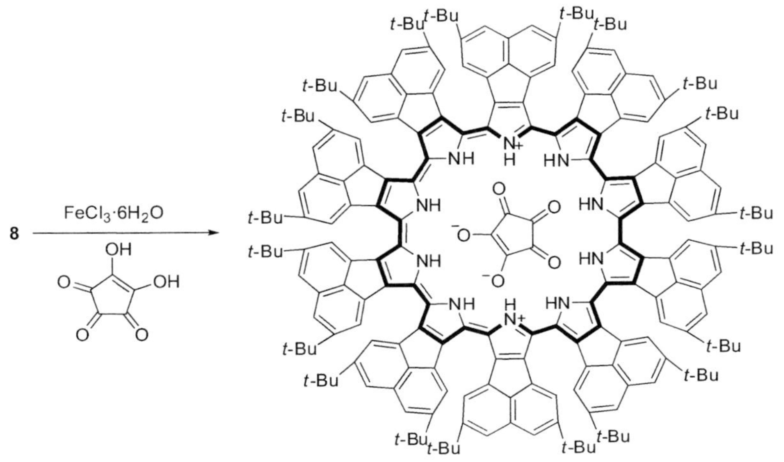

Scheme 6.9 Oxidative coupling of **8** in the presence of croconic acid.

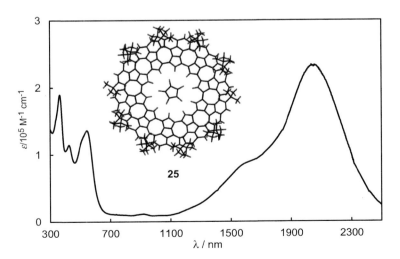

Figure 6.9 X-ray structure and UV-vis-NIR absorption spectrum of **25**.

6.6 Summary

π-Expanded and core-modified cyclo[n]pyrroles (n = 6~10), ring-expanded porphyrins with no *meso*-bridges have been successfully synthesized via oxidative coupling of 2,2'-bipyrroles with appropriate template anion and Suzuki–Miyaura coupling of terpyrrole analogues. The intense L band appears at 1000–2000 nm of NIR region. The wavelength of the absorption is controlled by the number of pyrroles and the number, positions, and types of peripheral fused-aromatic rings. The enhanced absorptions could prove useful for optelectronic and photofunctional devices in NIR dye applications. Challenging efforts are now underway to achieve selective synthesis of cyclo[n]pyrroles consisting of more than 10 pyrroles or a combination of pyrroles and other heterocycles in porphyrin chemistry.

Acknowledgments

This work was partially supported by JSPS KAKENHI (24750041 and 26410052).

References

1. Sessler, J. L. and Weghorn, S. J. (1997). *Expanded, Contracted & Isomeric Porphyrins*, Vol. 15 (Pergamon, New York).
2. Sessler, J. L., Gebauer, A., and Weghorn, S. J. (2000). Expanded porphyrins. In: *The Porphyrin Handbook*, Kadish, K. M., Smith, K. M., and Guilard, R. (Eds.) (Academic Press, San Diego), pp. 55–124.
3. Jasat, A. and Dolphin, D. (1997). Expanded porphyrins and their heterologs, *Chem. Rev.*, **97**, pp. 2267–2340.
4. Sessler, J. L. and Seidel, D. (2003). Synthetic expanded porphyrin chemistry, *Angew. Chem. Int. Ed.*, **42**, pp. 5134–5175.
5. Shimizu, S. and Osuka, A. (2006). Metalation chemistry of *meso*-aryl-substituted expanded porphyrins, *Eur. J. Inorg. Chem.*, pp. 1319–1335.
6. Yoon, Z. S., Osuka, A., and Kim, D. (2009). Möbius aromaticity and antiaromaticity in expanded porphyrins, *Nat. Chem.*, **1**, pp. 113–122.
7. Lim, J. M., Yoon, Z. S., Shin, J.-Y., Kim, K. S., Yoon, M.-C., and Kim, D. (2009). The photophysical properties of expanded porphyrins: Relationships

between aromaticity, molecular geometry and non-linear optical properties, *Chem. Commun.*, pp. 261–273.

8. Rambo, B. M. and Sessler, J. L. (2011). Oligopyrrole macrocycles: Receptors and chemosensors for potentially hazardous materials, *Chem. Eur. J.*, **17**, pp. 4946–4959.

9. Roznyatovskiy, V. V., Lee, C.-H., and Sessler, J. L. (2013). π-Extended isomeric and expanded porphyrins, *Chem. Soc. Rev.*, **42**, pp. 1921–1933.

10. Tanaka, T. and Osuka, A. (2017). Chemistry of *meso*-aryl-substituted expanded porphyrins: Aromaticity and molecular twist, *Chem. Rev.*, **117**, pp. 2584–2640.

11. Sarma, T. and Panda, P. K. (2017). Annulated isomeric, expanded, and contracted porphyrins, *Chem. Rev.*, **117**, pp. 2785–2838.

12. Shin, J.-Y., Furuta, H., Yoza, K., Igarashi, S., and Osuka, A. (2001). *meso*-Aryl-substituted expanded porphyrins, *J. Am. Chem. Soc.*, **123**, pp. 7190–7191.

13. Tanaka, Y., Saito, S., Mori, S., Aratani, N., Shinokubo, H., Shibata, N., Higuchi, Y., Yoon, Z. S., Kim, K. S., Noh, S. B., Park, J. K., Kim, D., and Osuka, A. (2008). Metalation of expanded porphyrins: A chemical trigger used to produce molecular twisting and Möbius aromaticity, *Angew. Chem. Int. Ed.*, **47**, pp. 681–684.

14. Seidel, D., Lynch, V., and Sessler, J. L. (2002). Cyclo[8]pyrrole: A simple-to-make expanded porphyrin with no meso bridges, *Angew. Chem. Int. Ed.*, **41**, pp. 1422–1425.

15. Eller, L. R., Stępień, M., Fowler, C. J., Lee, J. T., Sessler, J. L., and Moyer, B. A. (2007). Octamethyl-octaundecylcyclo[8]pyrrole: A promising sulfate anion extractant, *J. Am. Chem. Soc.*, **129**, pp. 11020–11021.

16. Sessler, J. L., Karnas, E., Kim, S. K., Ou, Z., Zhang, M., Kadish, K. M., Ohkubo, K., and Fukuzumi, S. (2008). "Umpolung" photoinduced charge separation in an anion-bound supramolecular complex, *J. Am. Chem. Soc.*, 130, pp. 15256–15257.

17. Stępień, M., Donnio, B., and Sessler, J. L. (2007). Supramolecular liquid crystals based on cyclo[8]pyrrole, *Angew. Chem. Int. Ed.*, **46**, pp. 1431–1435.

18. Lee, J. T., Chae, D.-H., Ou, Z., Kadish, K. M., Yao, Z., and Sessler, J. L. (2011). Unconventional Kondo effect in redox active single organic macrocyclic transistors, *J. Am. Chem. Soc.*, **133**, pp. 19547–19552.

19. Gorski, A., Köhler, T., Seidel, D., Lee, J. T., Orzanowska, G., Sessler, J. L., and Waluk, J. (2005). Electronic structure, spectra, and magnetic

circular dichroism of cyclohexa-, cyclohepta-, and cyclooctapyrrole, *Chem. Eur. J.*, **11**, pp. 4179–4184.

20. Alkorta, I., Blanco, F., and Elguero, J. (2009). A theoretical study of the neutral and the double-charged cation of cyclo[8]pyrrole and its interaction with inorganic anions, *Cent. Eur. J. Chem.*, **7**, pp. 683–689.

21. Lash, T. D. (2000). Synthesis of novel porphyrinoid chromophores. In: *The Porphyrin Handbook*, Kadish, K. M., Smith, K. M., and Guilard, R. (Eds.) (Academic Press, San Diego), pp. 125–199.

22. Ono, N., Yamada, H., and Okujima, T. (2010) Synthesis of porphyrins fused with aromatic rings. In: *Handbook of Porphyrin Science*, Kadish, K. M., Smith, K. M., and Guilard, R. (Eds.) (World Scientific, Singapore), pp. 1–102.

23. Okujima, T., Jin, G., Matsumoto, N., Mack, J., Mori, S., Ohara, K., Kuzuhara, D., Ando, C., Ono, N., Yamada, H., Uno, H., and Kobayashi, N. (2011). Cyclo[8]isoindoles: Ring-expanded and annelated porphyrinoids, *Angew. Chem. Int. Ed.*, **50**, pp. 5699–5703.

24. Roznyatovskiy, V. V., Lim, J. M., Lynch, V. M., Lee, B. S., Kim, D., and Sessler, J. L. (2011). π-Extension in expanded porphyrins: Cyclo[4]naphthobipyrrole, *Org. Lett.*, **13**, pp. 5620–5623.

25. Sarma, T. and Panda, P. K. (2011). Cyclo[4]naphthobipyrroles: Naphthobipyrrole-derived cyclo[8]pyrroles with strong near-infrared absorptions, *Chem. Eur. J.*, **17**, pp. 13987–13991.

26. Okujima, T., Ando, C., Mack, J., Mori, S., Hisaki, I., Nakae, T., Yamada, H., Ohara, K., Kobayashi, N., and Uno, H. (2013). Acenaphthylene-fused cyclo[8]pyrroles with intense near-IR-region absorption bands, *Chem. Eur. J.*, **19**, pp. 13970–13978.

27. Okujima, T., Ando, C., Mori, S., Nakae, T., Yamada, H., and Uno, H. (2014). Synthesis and molecular structure of cyclo[8](9,10-dihydro-9,10-anthraceno)pyrrole, *Heterocycles*, **88**, pp. 417–424.

28. Köhler, T., Seidel, D., Lynch, V., Arp, F. O., Ou, Z., Kadish, K. M., and Sessler, J. L. (2003). Formation and properties of cyclo[6]pyrrole and cyclo[7]pyrrole, *J. Am. Chem. Soc.*, **125**, pp. 6872–6873.

29. Okujima, T., Matsumoto, H., Mori, S., Nakae, T., Takase, M., and Uno, H. (2016). Synthesis of cyclo[8]pyrrole-polyoxometalate double-decker complex, *Tetrahedron Lett.*, **57**, pp. 3160–3162.

30. Harada, R., Matsuda, Y., Ōkawa, H., and Kojima, T. (2004). A porphyrin nanotube: Size-selective inclusion of tetranuclear molybdenum-oxo clusters, *Angew. Chem. Int. Ed.*, **43**, pp. 1825–1828.

31. Tsuda, A., Hirahara, E., Kim, Y.-S., Tanaka, H., Kawai, T., and Aida, T. (2004). A molybdenum crown cluster forms discrete inorganic–organic nanocomposites with metalloporphyrins, *Angew. Chem. Int. Ed.*, **43**, pp. 6327–6331.
32. Yokoyama, A., Kojima, T., Ohkubo, K., and Fukuzumi, S. (2007). A discrete conglomerate of a distorted Mo(V)-porphyrin with a directly coordinated Keggin-type polyoxometalate, *Chem. Commun.*, pp. 3997–3999.
33. Zhang, Z., Lim, J. M., Ishida, M., Roznyatovskiy, V. V., Lynch, V. M., Gong, H.-Y., Yang, X., Kim, D., and Sessler, J. L. (2012). Cyclo[*m*]pyridiene[*n*]pyrroles: Hybrid macrocycles that display expanded π-conjugation upon protonation, *J. Am. Chem. Soc.*, **134**, pp. 4076–4079.
34. Ho, I.-T., Zhang, Z., Ishida, M., Lynch, V. M., Cha, W.-Y., Sung, Y. M., Kim, D., and Sessler, J. L. (2014). A hybrid macrocycle with a pyridine subunit displays aromatic character upon uranyl cation complexation, *J. Am. Chem. Soc.*, **136**, pp. 4281–4286.
35. Bui, T.-T., Iordache, A., Chen, Z., Roznyatovskiy, V. V., Saint-Aman, E., Lim, J. M., Lee, B. S., Ghosh, S., Moutet, J.-C., Sessler, J. L., Kim, D., and Bucher, C. (2012). Electrochemical synthesis of a thiophene-containing cyclo[9]pyrrole, *Chem. Eur. J.*, **18**, pp. 5853–5859.
36. Zhang, Z., Cha, W.-Y., Williams, N. J., Rush, E. L., Ishida, M., Lynch, V. M., Kim, D., and Sessler, J. L. (2014). Cyclo[6]pyridine[6]pyrrole: A dynamic, twisted macrocycle with no *meso* bridges, *J. Am. Chem. Soc.*, **136**, pp. 7591–7594.
37. Okujima, T., Ando, C., Agrawal, S., Matsumoto, H., Mori, S., Ohara, K., Hisaki, I., Nakae, T., Takase, M., Uno, H., and Kobayashi, N. (2016). Template synthesis of decaphyrin without *meso*-bridges: Cyclo[10]pyrrole, *J. Am. Chem. Soc.*, **138**, pp. 7540–7543.

Chapter 7

Gold Isocyanide Complexes Exhibiting Luminescent Mechanochromism and Phase Transitions

Tomohiro Seki and Hajime Ito
Faculty of Engineering, Hokkaido University, Kita 13 Nishi 8 Kita-ku, Sapporo, Hokkaido, 060-8628, Japan
seki@eng.hokudai.ac.jp

7.1 Introduction

Luminescent mechanochromism is a phenomenon where emission colors of solid and liquid crystalline states of organic and organometallic compounds change upon mechanical stimulation such as grinding, pressing, shearing, rubbing, and ball-milling [1–5]. In these bulk materials, changes in assembled structures of molecules occur in response to mechanical stimulation concomitant with alteration of noncovalent interaction patterns between molecules. These alterations are responsible for changes in the photoluminescence property of mechanochromic compounds.

Light-Active Functional Organic Materials
Edited by Hiroko Yamada and Shiki Yagai
Copyright © 2019 Jenny Stanford Publishing Pte. Ltd.
ISBN 978-981-4800-15-0 (Hardcover), 978-0-429-44853-9 (eBook)
www.jennystanford.com

Such mechanochromic organic and organometallic compounds have potentials for applications in force sensors, probes, optical recording, displays, etc.

One of the pioneering examples of mechanochromic compounds was reported in 2005 [6, 7], and only around 10 reports have been known by 2009 [8–11]. After 2010, reports on luminescent mechanochromic compounds intensively increased and more than 500 reports contribute this research field so far [12–18]. Novel exploration concepts of mechanochromic compounds with new patterns of emission change, phase transition, and stimuli response are now required to expand this research field further. However, a critical problem to fulfill these requirements is difficulty in developing new mechanochromic compounds with desirable functionality. This is because designing molecular arrangements of small molecules in their bulk state, i.e., solid states, is significantly challenging for chemists at the present. Small molecules assemble in the bulk state through a variety of weak intermolecular interactions (π–π stacking, hydrogen bonds, CH–π interactions, ionic interactions, etc.), which are generally difficult to design and predict with high accuracy. For these reasons, present studies on luminescent mechanochromic compounds are largely on serendipitous discovery [19, 20].

Emission color control of mechanochromic compounds is also highly challenging subjects for exploration of mechanochromic compounds. The difficulty in controlling emission properties is due to the fact that emission properties of bulk materials are greatly influenced by molecular arrangements that cannot be precisely controlled as mentioned above. Emission colors of known luminescent mechanochromic compounds include blue, green, yellow, orange, and red, between which luminescent color change takes place upon exposure of mechanical stimulation. Among them, mechanochromic compounds possessing a red-emitting phase are scarce [21]. Especially, only a few reports of mechanochromic compounds exhibiting maximum emission wavelength ($\lambda_{em,max}$) of 700–800 nm have been known. For example, the difluoroboron compounds reported by Li and Zhang show a red emission ($\lambda_{em,max}$ = 752 nm) before exposure to mechanical stimulation [22], upon which their emission intensity decreases abruptly. The group of Enomoto has reported a luminescent mechanochromism of aminobenzopyranoxanthene, and it shows blue-shift from the emission band of $\lambda_{em,max}$ = 758 nm upon mechanical stimulation [23]. Li has reported a copper trinuclear complex showing a red emission band ($\lambda_{em,max}$ = 730 nm) after mechanical stimulation [24]. Mechanochromic compounds with IR emission after mechanical stress have not been reported so far.

As an approach to control emission properties of mechanochromic compounds, we recently tried to develop new compounds showing photoluminescence at longer wavelength region based on two different methods: systematic introduction of various substituents, and expansion of π-conjugated moieties of the chromophores. We found that these studies could provide important insights for the design of new mechano-responsive compounds. These will be the main topics of discussion in this chapter.

7.2 Emission Control of Aryl Gold Isocyanide Complexes

7.2.1 Aryl Gold Isocyanide Complexes

Our group reported a luminescent mechanochromic gold isocyanide complex **A** (Fig. 7.1) in 2008 [25]. Blue-emitting pristine semi-crystals of **A** transformed to yellow-emitting powder upon mechanical stimulation (Fig. 7.2a). Remarkably, the addition of a solvent such as CH_2Cl_2 to ground powder, the original blue emission is recovered (Fig. 7.2a). At that time, only several mechanochromic compounds have been known as described above and complex **A** is the first example of reversible mechanochromic compound [25].

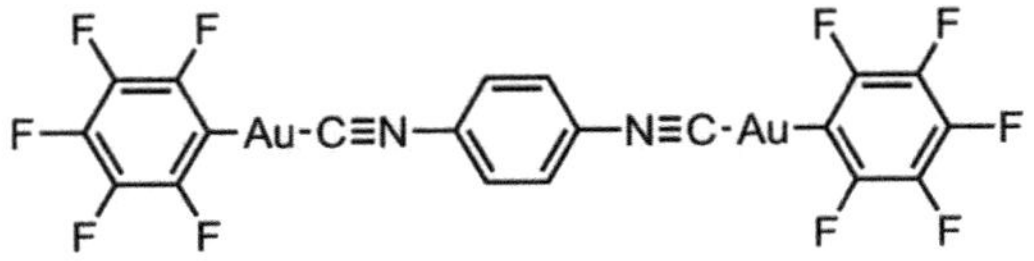

Figure 7.1 Molecular structure of complex **A**.

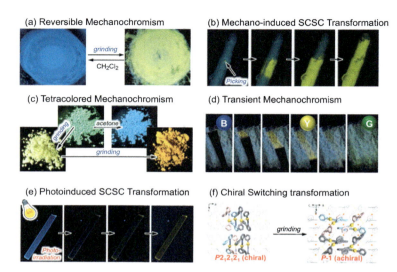

Figure 7.2 Overview of unique mechanochromism of several aryl gold isocyanide complexes reported by our group [25, 27–35].

Since the seminal report on complex **A**, we have intensively studied luminescent mechanochromism by creating a series of structurally related aryl gold isocyanide complexes (Fig. 7.2). Before our intensive studies, only a limited synthetic procedure for aryl gold isocyanide complexes was known [26] and their mechanochromic properties have never been reported. Our research so far revealed a series of intriguing luminescent mechanochromic properties of aryl gold isocyanide complexes by developing a new synthetic procedure for them. For example, we reported the first two examples of luminescent mechanochromism based on single-crystal-to-single-crystal (SCSC) phase transitions, and this is referred to as mechano-induced SCSC phase transition (Fig. 7.2b) [27–29]. Because mechanical stimulation is random and rough stimuli, phase transition while keeping crystallinity and macroscopic crystal integrity is rare. We also reported several types of "multiple" colored luminescent mechanochromism: One shows interconvertible tetra-colored photoluminescence (Fig. 7.2c) [30] and others show stepwise emission color changes from blue to yellow and then to green only by one scratching (Fig. 7.2d) [31,

32]. We also found that another mechanochromic gold isocyanide complex showed both photoinduced phase transition and mechano-induced phase transition (Fig. 7.2e). As the first example, we report a gold isocyanide showing the phase transition that can be induced only by mechanical stimulation or photoirradiation but not induced by thermal treatment [33, 34]. Phase transition from chiral to non-chiral space groups upon luminescent mechanochromism of a gold isocyanide complex with biphenyl moiety was also reported as the first example (Fig. 7.2f) [35]. We also investigated the semi-rational control of emission colors of our aryl gold isocyanide complexes, as shown in the next section and Table 7.1 in detail.

Table 7.1 Structure of R^1–R^2 complexes

R^1\R^2	NMe₂	OMe	Me	H	Cl	CF₃	CN	NO₂
OMe								
Me								
H								
Cl								
CF₃								
CN								

R^1—⟨phenyl⟩—Au-C≡N—⟨phenyl⟩—R^2
$R^1 - R^2$

(left panel) before grinding
(right panel) after grinding

Note: Photos of the powder samples of R^1–R^2 complexes before and after mechanical stimulation taken under UV light.

7.2.2 Substituents Effect

In 2016, in order to control the emission wavelength of aryl gold isocyanide complexes exhibiting luminescent mechanochromism, we systematically prepared a series of gold isocyanide complexes possessing various substituents [36]. We decided phenyl(phenyl isocyanide)gold complexes [29] as the main scaffold structure and introduced various electron-withdrawing/donating substituents into two *para*-positions of their terminal benzene rings. We selected six kinds in R^1 moieties (OMe, Me, H, Cl, CF$_3$, and CN) and eight kinds in R^2 moieties (NMe$_2$, OMe, Me, H, Cl, CF$_3$, CN, and NO$_2$). Therefore, we prepared 48 **R^1-R^2** complexes. Table 7.1 shows photographs of the powder samples of 48 **R^1-R^2** complexes taken under UV light. Each panel corresponds to one of **R^1-R^2** complexes. Each panel has two powder samples: left-side sample corresponds to the powder before mechanical stimulation, and right-side sample corresponds to the powder after mechanical stimulation.

After preparing 48 **R^1-R^2** complexes, we adopted a "screening approach" for finding new mechanochromic compounds. We found a complex that showed a unique type of mechanochromism in which changes in its molecular arrangement upon grinding were based on crystal-to-crystal phase transition, rather than the more common crystal-to-amorphous phase transition [36]. In our "screening approach," we performed three-step experiments. The first screening step was the selection of emissive complexes under UV light, affording 37 emissive **R^1-R^2** complexes with photoluminescence. For these 37 complexes, we performed the second screening and investigated the mechanochromic properties through emission spectroscopy. Among these 37 complexes, 28 complexes showed mechanochromic property. In the third screening step, we evaluated the remaining complexes for changes in their powder diffraction patterns induced by mechanical stimulation. As a result, we newly found that the **CF$_3$-CN** complex showed mechanochromism with crystal-to-crystal phase transition [36]. In the literature, only less than 10% of mechanochromic compounds showed crystal-to-crystal phase transition. This is one rare example.

When we focused on the emission colors of 48 **R^1-R^2** complexes (Table 7.1), we found a certain degree of correlations between $\lambda_{em,max}$ and electron-withdrawing/donating characters. We found that the $\lambda_{em,max}$ of **R^1-R^2** complexes (both unground and ground phases) shifted to longer wavelength region when the electron-withdrawing character of R^2 moieties was increased (from left to right side in Table 7.1). In other words, blue and green emissions are often found in the left side of Table 7.1, and yellow and orange emission colors are mainly found in the right side. This result indicates that Table 7.1 can be used for a reference library in which the desired emission colors and changing patterns can be selected.

The group of You also succeeded in preparing mechanochromic compounds exhibiting various emission colors through systematic molecular synthesis by taking advantage of the C–H activation process [37]. However, they found that simple alternation of the substituents of the main mechanochromic chromophores was not very suitable for achieving the emission properties in further longer wavelength regions, such as deep-red and infrared regions.

7.3 IR Emissive Mechanochromic Compounds

7.3.1 Extension of π-Conjugated Systems

In order to develop mechanochromic compounds exhibiting emission at longer wavelength regions, π-conjugated aryl moieties were extended. Figure 7.3 (top-left) shows the molecular structure of the **H–H** complex (Table 7.1), which will be referred to as complex **1** hereafter. A phenyl group on the Au atom of **1** was expanded to the naphthyl group in complex **2**. As shown in Fig. 7.3, complex **2** also showed luminescent mechanochromism upon grinding in which the as-prepared form of **2**, **2a** with green emission transformed to ground form **2b** with orange emission. This emission color is clearly different from that of complex **1**. The emission spectrum of **2a** showed three peaks, with the maximum at 523 nm (Fig. 7.3). On the other hand, orange-emitting ground powder **2b** showed a broad emission spectrum with a maximum at 599 nm. This $\lambda_{em,max}$ of **2b** is longer than those of all **R¹–R²** complexes (Table 7.1) among which the ground powder of the **Me–CN** complex showed the longest emission with maximum wavelength at 592 nm [36]. This result indicates that the extension of π-conjugated moieties is suitable for long-wavelength emission of mechanochromic materials. Indeed, complex **3** containing more extended anthracene showed remarkably long-wavelength emission even in the infrared region [38].

140 | Gold Isocyanide Complexes Exhibiting Luminescent Mechanochromism

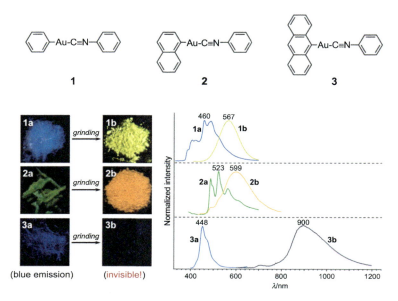

Figure 7.3 Structures of **1**, **2**, and **3** (top). Photos (bottom left, taken under UV light) and emission spectra (bottom right, excitation at 365 m) of the powder samples before and after applying mechanical stimulation of **1**, **2**, and **3**.

7.3.2 Anthryl Gold Isocyanide Complexes

The anthryl gold isocyanide complex **3** was prepared according to a previously reported procedure [29]. The as-prepared solid **3** (hereafter denoted as **3a**) is a semi-crystalline sample exhibiting blue emission under excitation at 365 nm. Complex **3a** shows emission spectrum peaked at 448 nm (Fig. 7.3). Unexpectedly, **3a** showed $\lambda_{em,max}$ at a shorter wavelength compared to those of unground phases of **1** and **2**, less π-extended analogues. It exhibited absolute emission quantum yield (Φ_{em}) and average emission lifetime (τ_{av}) [= $\Sigma(A_n\tau_n^2)/\Sigma(A_n\tau_n)$] values of 0.5% and 0.11 ns, respectively. Interestingly, τ_{av} (0.11 ns) of **3a** indicates that the emission mechanism is based on fluorescence. These are highly contrast to those of other structurally related aryl gold isocyanide complexes [25, 27–36] in which the emissions of both unground and ground samples are based on phosphorescence.

Complex **3a** shows unprecedented luminescent mechanochromism upon grinding. After grinding **3a**, we obtained ground powder **3b**, from which visible photoluminescence could not be observed (Fig. 7.3). However, **3b** exhibited broad emission bands in the IR region of 800–1200 nm with $\lambda_{em,max}$ at 900 nm (Fig. 7.3). This $\lambda_{em,max}$ value is the longest wavelength emission for luminescent mechanochromic compounds reported so far. The magnitude of the spectral shift for **3a** upon mechanical stimulation to form **3b** is, to the best of our knowledge, unprecedented, i.e., $\Delta\lambda_{em,max}$ = 452 nm (1.39 eV). Powder **3b** exhibits Φ_{em} of 0.09%. Generally, the Φ_{em} value of the solid-state IR emission ($\lambda_{em,max}$ > 800 nm) is very low because efficient nonradiative relaxation can occur. For example, the group of Fages has reported that a boron difluoride complex shows $\lambda_{em,max}$ = 855 nm with Φ_{em} = 0.1% in the solid state [39]. Powder **3b** shows τ_{av} = 0.69 µs, indicating phosphorescence character typical for gold complexes [25, 27–36]. This indicates that luminescent mechanochromism of **3a** to form **3b** is accompanied by a change in the emission mechanism from fluorescence to phosphorescence. Such fluorescence/phosphorescence switching is rarely observed for mechanochromic compounds [24].

In order to obtain insight into the mechanism of luminescent mechanochromism of **3**, we performed X-ray diffraction analyses. Crystallization of **3** from CH_2Cl_2 and MeOH afforded single crystals belonging to a $P2_1/n$ space group (Fig. 7.4). As shown in Fig. 7.4a, the dimer unit adopts a head-to-tail arrangement with two prominent CH/π interactions between the anthracene ring and a hydrogen atom of the phenyl ring ($L_{CH/\pi}$ = 2.778 Å). The molecules take twisted conformation: The dihedral angle between the anthracene plane and the phenyl ring θ is 86.17°, indicating the almost perpendicular conformation. We could not confirm a prominent intermolecular interaction between the dimers (Fig. 7.4b). Entire molecules are packed in a herringbone structure (Fig. 7.4c). Based on the single-crystal coordination of **3a**, we performed time-dependent (TD) DFT calculations. The result revealed that a twisted conformation of **3a** ($\theta \approx 90°$) caused a forbidden ligand-to-ligand charge transfer (LLCT) as the transition from the lowest excited state, i.e., the transition from the highest occupied molecular orbital (HOMO) (localized at the anthracene moiety) to the lowest unoccupied molecular orbital (LUMO) (localized at the AuCNPh moiety) was not possible in this conformation of molecules. Instead, the emission of **3a** originates from the HOMO→LUMO+1 transition (π–π^* transition) in which both participant orbitals are localized on the anthracene moiety of **3a**. This is in line with the unexpected experimental results of **3a**: short wavelength emission ($\lambda_{em,max}$ = 448 nm) with short emission lifetime (τ_{av} = 0.11 ns, fluorescent).

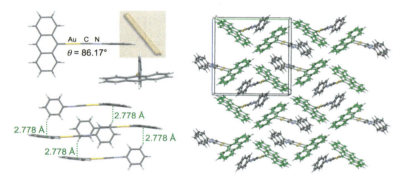

Figure 7.4 Single-crystal structure of **3a**.

We next investigated the relationships between the solid structures and the emission mechanisms of the ground powder of **3b** by means of powder X-ray diffraction. The pristine sample **3a** showed intense diffraction peaks (Fig. 7.5), again confirming its crystallinity. This diffraction pattern of **3a** also matched with simulated powder pattern derived from the single-crystal samples. Complex **3b** exhibited the featureless diffraction pattern (Fig. 7.5), indicating the amorphous structure of **3b**. The X-ray diffraction studies indicated that crystal-to-amorphous phase transition took place during the mechanochromism of **3**. This is the typical molecular arrangement changes found in many mechanical responsive materials [36].

We further investigated the emission mechanism of amorphous phase **3b** by IR absorption spectroscopy. It was revealed that the IR absorption of the C≡N stretching vibrations of gold isocyanide complexes in solid phases shifted to a lower wavenumber when aurophilic interactions were involved [25, 40]. The IR studies of our previously reported four aryl gold isocyanide complexes with related structures [28, 29, 33, 36] support that the shift in the C≡N stretching vibration to lower wavenumbers could be an indicator of the formation of aurophilic interactions. For **3b**, the absorption peak for the C≡N stretching vibrations is observed at 2193 cm^{-1}. Upon grinding, the C≡N stretching mode of **3a** at 2200 cm^{-1} thus shifts to lower wavenumbers. This indicates the presence of aurophilic interactions in the ground powder **3b**. It is well known that the formation of aurophilic interaction usually causes red-shifted emission band in the photoluminescence of gold complexes [41–44] (Fig. 7.6). In addition, it has been reported that mechanochromic anthracene molecules form π–π stacking interaction in their amorphous phases [45, 46]. For **3b**, combination of aurophilic interactions and π–π stacking interactions formed in the amorphous phase is believed to be the key to achieve unprecedented IR emission after mechanical stimulation.

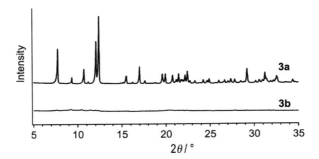

Figure 7.5 Powder X-ray diffraction patterns of **3a** and **3b**.

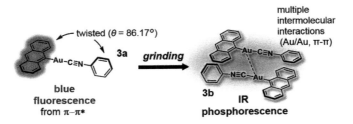

Figure 7.6 Schematic representation of emission mechanism of unground **3a** and ground phase **3b**.

7.4 Summary

This chapter elaborates our recent studies on luminescent mechanochromic gold isocyanide complexes. Control over the luminescent colors of mechanochromic compounds is considered to be challenging owing to the difficulty in the construction of the desired solid-state structures of compounds. As a demonstration of controllable emission property, we have studied mechanochromic compounds with long-wavelength emissions by using a novel mechanochromic scaffold: aryl gold isocyanide complexes. The results indicate that extension of π-conjugated moieties is more promising compared to the introduction of electron-withdrawing/donating substituents to express photoluminescence at long-wavelength regions. Moreover, we have found that anthracene-based gold isocyanide complex **3** can exhibit unprecedented IR emission ($\lambda_{em,max}$ = 900 nm) in response to mechanical stimulation. This is the first compound that shows both solid-state IR emission and luminescent mechanochromism. Generally, luminescent mechanochromic compounds would be applied to force sensing and memory devices because of the mechano-induced emission color changes within the visible region. In addition, the newly prepared mechanochromic compounds with IR emission have a potential to be applied to bioimaging and security inks by taking advantage of their photoluminescence that is invisible to naked eyes.

Acknowledgments

This work was financially supported by the Program for Fostering Researchers for the Next Generation and by the JSPS KAKENHI grants JP16H06034, JP17H06370, JP17H05134, and JP17H05344. We also thank the Frontier Chemistry Center Akira Suzuki "Laboratories for Future Creation" Project, Hokkaido University, for providing us with access to equipment.

References

1. Balch, A. L. (2009) Dynamic crystals: Visually detected mechanochemical changes in the luminescence of gold and other transition-metal complexes, *Angew. Chem. Int. Ed.*, **48**, pp. 2641–2644.

2. Sagara, Y. and Kato, T. (2009) Mechanically induced luminescence changes in molecular assemblies, *Nat. Chem.*, **1**, pp. 605–610.
3. Sagara, Y., Yamane, S., Mitani, M., Weder, C., and Kato, T. (2016) Mechanoresponsive luminescent molecular assemblies: An emerging class of materials, *Adv. Mater.*, **28**, pp. 1073–1095.
4. Seki, T. and Ito, H. (2016) Molecular-level understanding of structural changes of organic crystals induced by macroscopic mechanical stimulation, *Chem. Eur. J.*, **22**, pp. 4322–4329.
5. Chi, Z., Zhang, X., Xu, B., Zhou, X., Ma, C., Zhang, Y., Liu, S., and Xu, J. (2012) Recent advances in organic mechanofluorochromic materials, *Chem. Soc. Rev.*, **41**, pp. 3878–3896.
6. Lee, Y. A. and Eisenberg, R. (2003) Luminescence tribochromism and bright emission in gold(I) thiouracilate complexes, *J. Am. Chem. Soc.*, **125**, pp. 7778–7779.
7. Mizukami, S., Houjou, H., Sugaya, K., Koyama, E., Tokuhisa, H., Sasaki, T., and Kanesato, M. (2005) Fluorescence color modulation by intramolecular and intermolecular π–π interactions in a helical zinc(II) complex, *Chem. Mater.*, **17**, pp. 50–56.
8. Sagara, Y., Mutai, T., Yoshikawa, I., and Araki, K. (2007) Material design for piezochromic luminescence: Hydrogen-bond-directed assemblies of a pyrene derivative, *J. Am. Chem. Soc.*, **129**, pp. 1520–1521.
9. Kunzelman, J., Kinami, M., Crenshaw, B. R., Protasiewicz, J. D., and Weder, C. (2008) Oligo(*p*-phenylene vinylene)s as a "new" class of piezochromic fluorophores, *Adv. Mater.*, **20**, pp. 119–122.
10. Ooyama, Y., Kagawa, Y., Fukuoka, H., Ito, G., and Harima, Y. (2009) Mechanofluorochromism of a series of benzofuro[2,3-*c*]oxazolo[4,5-*a*] carbazole-type fluorescent dyes, *Eur. J. Org. Chem.*, **2009**, pp. 5321–5326.
11. Chung, J. W., You, Y., Huh, H. S., An, B. K., Yoon, S. J., Kim, S. H., Lee, S. W., and Park, S. Y. (2009) Shear- and UV-induced fluorescence switching in stilbenic π-dimer crystals powered by reversible [2 + 2] cycloaddition, *J. Am. Chem. Soc.*, **131**, pp. 8163–8172.
12. Yoon, S.-J., Chung, J. W., Gierschner, J., Kim, K. S., Choi, M.-G., Kim, D., and Park, S. Y. (2010) Multistimuli two-color luminescence switching *via* different slip-stacking of highly fluorescent molecular sheets, *J. Am. Chem. Soc.*, **132**, pp. 13675–13683.
13. Zhang, G., Lu, J., Sabat, M., and Fraser, C. L. (2010) Polymorphism and reversible mechanochromic luminescence for solid-state difluoroboron avobenzone, *J. Am. Chem. Soc.*, **132**, pp. 2160–2162.

14. Nagura, K., Saito, S., Yusa, H., Yamawaki, H., Fujihisa, H., Sato, H., Shimoikeda, Y., and Yamaguchi, S. (2013) Distinct responses to mechanical grinding and hydrostatic pressure in luminescent chromism of tetrathiazolylthiophene, *J. Am. Chem. Soc.*, **135**, pp. 10322–10325.

15. Sagara, Y., Komatsu, T., Ueno, T., Hanaoka, K., Kato, T., and Nagano, T. (2014) Covalent attachment of mechanoresponsive luminescent micelles to glasses and polymers in aqueous conditions, *J. Am. Chem. Soc.*, **136**, pp. 4273–4280.

16. Yagai, S., Okamura, S., Nakano, Y., Yamauchi, M., Kishikawa, K., Karatsu, T., Kitamura, A., Ueno, A., Kuzuhara, D., Yamada, H., Seki, T., and Ito, H. (2014) Design amphiphilic dipolar π-systems for stimuli-responsive luminescent materials using metastable states, *Nat. Commun.*, **5**, pp. 4013.

17. Ito, S., Taguchi, T., Yamada, T., Ubukata, T., Yamaguchi, Y., and Asami, M. (2017) Indolylbenzothiadiazoles with varying substituents on the indole ring: A systematic study on the self-recovering mechanochromic luminescence, *RSC Adv.*, **7**, pp. 16953–16962.

18. Sagara, Y., Kubo, K., Nakamura, T., Tamaoki, N., and Weder, C. (2017) Temperature-dependent mechanochromic behavior of mechanoresponsive luminescent compounds, *Chem. Mater.*, **29**, pp. 1273–1278.

19. Li, G., Song, F., Wu, D., Lan, J., Liu, X., Wu, J., Yang, S., Xiao, D., and You, J. (2014) Cation-anion interaction-directed molecular design strategy for mechanochromic luminescence, *Adv. Funct. Mater.*, **24**, pp. 747–753.

20. Han, T., Zhang, Y., Feng, X., Lin, Z., Tong, B., Shi, J., Zhi, J., and Dong, Y. (2013) Reversible and hydrogen bonding-assisted piezochromic luminescence for solid-state tetraaryl-buta-1,3-diene, *Chem. Commun.*, **49**, pp. 7049–7051.

21. Kwon, M. S., Gierschner, J., Yoon, S. J., and Park, S. Y. (2012) Unique piezochromic fluorescence behavior of dicyanodistyrylbenzene based donor-acceptor-donor triad: Mechanically controlled photo-induced electron transfer (eT) in molecular assemblies, *Adv. Mater.*, **24**, pp. 5487–5492.

22. Cheng, X., Li, D., Zhang, Z., Zhang, H., and Wang, Y. (2014) Organoboron compounds with morphology-dependent NIR emissions and dual-channel fluorescent ON/OFF switching, *Org. Lett.*, **16**, pp. 880–883.

23. Tanioka, M., Kamino, S., Muranaka, A., Ooyama, Y., Ota, H., Shirasaki, Y., Horigome, J., Ueda, M., Uchiyama, M., Sawada, D., and Enomoto, S.

(2015) Reversible near-infrared/blue mechanofluorochromism of aminobenzopyranoxanthene, *J. Am. Chem. Soc.*, **137**, pp. 6436–6439.
24. Xiao, Q., Zheng, J., Li, M., Zhan, S. Z., Wang, J. H., and Li, D. (2014) Mechanically triggered fluorescence/phosphorescence switching in the excimers of planar trinuclear copper(I) pyrazolate complexes, *Inorg. Chem.*, **53**, pp. 11604–11615.
25. Ito, H., Saito, T., Oshima, N., Kitamura, N., Ishizaka, S., Hinatsu, Y., Wakeshima, M., Kato, M., Tsuge, K., and Sawamura, M. (2008) Reversible mechanochromic luminescence of $[(C_6F_5Au)_2(\mu\text{-}1,4\text{-diisocyanobenzene})]$, *J. Am. Chem. Soc.*, **130**, pp. 10044–10045.
26. Bayon, R., Coco, S., Espinet, P., Fernandez-Mayordomo, C., and Martin-Alvarez, J. M. (1997) Liquid-crystalline mono- and dinuclear (perhalophenyl)gold(I) isocyanide complexes, *Inorg. Chem.*, **36**, pp. 2329–2334.
27. Seki, T., Sakurada, K., Muromoto, M., Seki, S., and Ito, H. (2016) Detailed investigation of the structural, thermal, and electronic properties of gold isocyanide complexes with mechano-triggered single-crystal-to-single-crystal phase transitions, *Chem. Eur. J.*, **22**, pp. 1968–1978.
28. Seki, T., Sakurada, K., and Ito, H. (2013) Controlling mechano- and seeding-triggered single-crystal-to-single-crystal phase transition: Molecular domino with a disconnection of aurophilic bonds, *Angew. Chem. Int. Ed.*, **52**, pp. 12828–12832.
29. Ito, H., Muromoto, M., Kurenuma, S., Ishizaka, S., Kitamura, N., Sato, H., and Seki, T. (2013) Mechanical stimulation and solid seeding trigger single-crystal-to-single-crystal molecular domino transformations, *Nat. Commun.*, **4**, article number 2009.
30. Seki, T., Ozaki, T., Okura, T., Asakura, K., Sakon, A., Uekusa, H., and Ito, H. (2015) Interconvertible multiple photoluminescence color of a gold(I) isocyanide complex in the solid state: Solvent-induced blue-shifted and mechano-responsive red-shifted photoluminescence, *Chem. Sci.*, **6**, pp. 2187–2195.
31. Yagai, S., Seki, T., Aonuma, H., Kawaguchi, K., Karatsu, T., Okura, T., Sakon, A., Uekusa, H., and Ito, H. (2016) Mechanochromic luminescence based on crystal-to-crystal transformation mediated by a transient amorphous state, *Chem. Mater.*, **28**, pp. 234–241.
32. Seki, T., Kashiyama, K., Yagai, S., and Ito, H. (2017) Tuning the lifetime of transient phases of mechanochromic gold isocyanide complexes through functionalization of the terminal moieties of flexible side chains, *Chem. Lett.*, **46**, pp. 1415–1418.

33. Seki, T., Sakurada, K., Muromoto, M., and Ito, H. (2015) Photoinduced single-crystal-to-single-crystal phase transition and photosalient effect of a gold(I) isocyanide complex with shortening of intermolecular aurophilic bonds, *Chem. Sci.*, **6**, pp. 1491–1497.

34. Sakurada, K., Seki, T., and Ito, H. (2016) Mechanical path to a photogenerated structure: Ball milling-induced phase transition of a gold(I) complex, *CrystEngComm*, **18**, pp. 7217–7220.

35. Jin, M., Seki, T., and Ito, H. (2017) Mechano-responsive luminescence via crystal-to-crystal phase transitions between chiral and non-chiral space groups, *J. Am. Chem. Soc.*, **139**, pp. 7452–7455.

36. Seki, T., Takamatsu, Y., and Ito, H. (2016) A screening approach for the discovery of mechanochromic gold(I) isocyanide complexes with crystal-to-crystal phase transitions, *J. Am. Chem. Soc.*, **138**, pp. 6252–6260.

37. Wu, J., Cheng, Y., Lan, J., Wu, D., Qian, S., Yan, L., He, Z., Li, X., Wang, K., Zou, B., and You, J. (2016) Molecular engineering of mechanochromic materials by programmed C-H arylation: Making a counterpoint in the chromism trend, *J. Am. Chem. Soc.*, **138**, pp. 12803–12812.

38. Seki, T., Tokodai, N., Omagari, S., Nakanishi, T., Hasegawa, Y., Iwasa, T., Taketsugu, T., and Ito, H. (2017) Luminescent mechanochromic 9-anthryl gold(I) isocyanide complex with an emission maximum at 900 nm after mechanical stimulation, *J. Am. Chem. Soc.*, **139**, pp. 6514–6517.

39. D'Aleo, A., Gachet, D., Heresanu, V., Giorgi, M., and Fages, F. (2012) Efficient NIR-light emission from solid-state complexes of boron difluoride with 2'-hydroxychalcone derivatives, *Chem. Eur. J.*, **18**, pp. 12764–12772.

40. White-Morris, R. L., Olmstead, M. M., and Balch, A. L. (2003) Aurophilic interactions in cationic gold complexes with two isocyanide ligands. Polymorphic yellow and colorless forms of [(cyclohexyl isocyanide)$_2$AuI](PF$_6$) with distinct luminescence, *J. Am. Chem. Soc.*, **125**, pp. 1033–1040.

41. Balch, A. L. (2004) Polymorphism and luminescent behavior of linear, two-coordinate gold(I) complexes, *Gold Bull.*, **37**, pp. 45–50.

42. Pyykkö, P. (2004) Theoretical chemistry of gold, *Angew. Chem. Int. Ed.*, **43**, pp. 4412–4456.

43. Schmidbaur, H. and Schier, A. (2008) A briefing on aurophilicity, *Chem. Soc. Rev.*, **37**, pp. 1931–1951.

44. He, X. and Yam, V. W.-W. (2011) Luminescent gold(I) complexes for chemosensing, *Coord. Chem. Rev.*, **255**, pp. 2111–2123.
45. Teng, M., Wang, Z., Ma, Z., Chen, X., and Jia, X. (2014) Mechanochromic and photochromic dual-responsive properties of an amino acid based molecule in polymorphic phase, *RSC Adv.*, **4**, pp. 20239–20241.
46. Dong, Y., Xu, B., Zhang, J., Tan, X., Wang, L., Chen, J., Lv, H., Wen, S., Li, B., Ye, L., Zou, B., and Tian, W. (2012) Piezochromic luminescence based on the molecular aggregation of 9,10-bis((*E*)-2-(pyrid-2-yl)vinyl) anthracene, *Angew. Chem. Int. Ed.*, **51**, pp. 10782–10785.

Chapter 8

Thermally Activated Delayed Fluorescence Materials for Organic Light-Emitting Devices

Kyohei Matsuo, Naoya Aizawa, and Takuma Yasuda
INAMORI Frontier Research Center (IFRC), Kyushu University,
744 Motooka, Nishi-ku, Fukuoka 818-0395, Japan
yasuda@ifrc.kyushu-u.ac.jp

8.1 Introduction

The design and synthesis of highly luminescent organic molecules are key issues for various optoelectronic applications such as organic light-emitting diodes (OLEDs), organic light-emitting transistors, and organic lasers. Since the first observation of electroluminescence (EL) from organic crystals in 1963 [1], a number of emitter molecules have been developed to enhance the EL efficiency. In general, the EL efficiency is evaluated with the external EL quantum efficiency (η_{ext}), which corresponds to the number of photons emitted from the device to air per charge carriers injected into the device. η_{ext} is expressed by Eq. (8.1):

Light-Active Functional Organic Materials
Edited by Hiroko Yamada and Shiki Yagai
Copyright © 2019 Jenny Stanford Publishing Pte. Ltd.
ISBN 978-981-4800-15-0 (Hardcover), 978-0-429-44853-9 (eBook)
www.jennystanford.com

$$\eta_{\text{ext}} = \eta_{\text{int}} \cdot \eta_{\text{out}} = (\gamma \cdot \eta_{\text{ST}} \cdot \Phi_{\text{PL}}) \eta_{\text{out}} \qquad (8.1)$$

where η_{int} is the internal EL quantum efficiency, η_{out} is the light out-coupling efficiency, γ is the charge balance factor, η_{ST} is the fraction of radiative excitons, and Φ_{PL} is the photoluminescence (PL) quantum yield of the emitter. γ and Φ_{PL} can be enhanced to nearly 100% by adopting the advanced device architectures and sophisticated molecular design of emitters, respectively. Typical η_{out} is approximately 20% if the emitters are randomly oriented in the emission layer without any out-coupling enhancement. More importantly, η_{ST} depends on the EL mechanisms and is typically limited to 25% for conventional fluorescent emitters. As a result, the theoretical maximum of η_{ext} is limited to 5% for fluorescent OLEDs.

Recently, thermally activated delayed fluorescence (TADF) materials have attracted a great deal of attention as a new generation of emitter for OLEDs because they can overcome the limitation of theoretical maxima of η_{ext} of 5%, without using expensive precious metals. Since the first OLED application of a purely organic TADF emitter in 2011 [2], considerable efforts have been devoted to exploring the molecular design of efficient TADF materials and understanding structure–property relationship both experimentally and theoretically.

This chapter focuses on the molecular design and optoelectronic properties of purely organic TADF materials for OLEDs. First, some basic principles for TADF-OLEDs are described. Then the molecular design concepts for efficient TADF materials are demonstrated. Finally, some recent advances in more functionalized TADF materials are also described.

8.2 Basic Principles of TADF-OLEDs

8.2.1 EL Mechanism in OLEDs

OLEDs have a multilayer structure, in which thin films of organic semiconductors are sandwiched between two electrodes (Fig. 8.1). In OLEDs, the holes and electrons are injected by applying an external electrical field from the anode and cathode, respectively. Upon recombination of the holes and electrons in the emission layer, the excitons of emitter molecules are electrochemically generated.

While only singlet excited states are generated in photoexcitation from singlet ground (S_0) state, electrical excitation typically leads to the formation of singlet and triplet excitons with a probability of 25% and 75%, respectively, according to spin statistics [3]. Therefore, the theoretical maximum of η_{int} is limited to 25% for OLEDs using conventional fluorescent emitters, because they can utilize only the singlet excitons for light emission and triplet excitons are deactivated through non-radiative decay pathways (Fig. 8.2a). To overcome this limitation, employing the triplet excitons for EL is essential. One of the most effective ways is to utilize phosphorescence from heavy transition metal complexes such as tris(2-phenylpyridinato) iridium(III) [4] (Fig. 8.2b). Although phosphorescence, i.e., the radiative transition from the lowest excited triplet (T_1) states to the S_0 state, is a spin-forbidden transition process, the strong spin–orbit coupling of heavy metals facilitates the transition rate. In addition, intersystem crossing (ISC) from the lowest excited singlet (S_1) state to the T_1 state is also enhanced by heavy metals and, hence, the singlet excitons are rapidly converted to the triplet excitons. As a result, all electro-generated excitons can be utilized for light emission, rendering η_{int} of nearly 100%. However, highly emissive phosphorescent materials are limited to heavy metal complexes, which include expensive precious metals such as Ir and Pt.

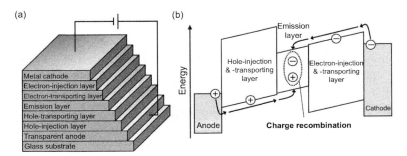

Figure 8.1 (a) Device architecture of typical OLEDs. (b) Schematic energy diagram of OLEDs.

As a new alternative way for harvesting the triplet excitons and producing efficient OLEDs, the use of TADF has been proposed (Fig. 8.2c). In this TADF mechanism, reverse intersystem crossing (RISC) from the T_1 state to the S_1 state is a key process. Since the

T_1 state is energetically lower than the S_1 state, RISC involves an endothermic process. When the singlet–triplet energy splitting (ΔE_{ST}) is sufficiently small, RISC can be activated by thermal energy ($k_B T \approx 25.6$ meV) at room temperature. Consequently, the non-emissive triplet excitons are up-converted to the emissive singlet excitons, leading to a high EL efficiency. An important advantage of TADF materials is that they do not require heavy metals, which is beneficial to reducing the cost of device fabrication.

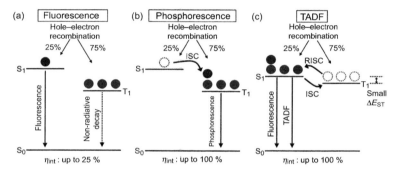

Figure 8.2 Schematic representation of three different types of emission mechanism in OLEDs: (a) fluorescence, (b) phosphorescence, and (c) TADF.

8.2.2 Basic Principles for TADF

RISC and ISC are generally spin-forbidden transitions for most luminescent organic molecules. Thus, they require mixing of the singlet and triplet states. Spin–orbit coupling (H_{SO}) causes the mixing of the excited state, and those transitions become allowed to a certain extent. According to perturbation theory, the first-order mixing coefficient (λ) is expressed by Eq. (8.2) [5]:

$$\lambda = \left| \frac{<S_1|H_{SO}|T_1>}{\Delta E_{ST}} \right| \tag{8.2}$$

where $<S_1|H_{SO}|T_1>$ is the spin–orbit coupling matrix element. The spin–orbit coupling of purely organic molecules is much smaller than that of heavy metal complexes; minimizing the ΔE_{ST} is, thus, the most effective way to increase λ and enhance RISC in the design of TADF materials.

The TADF phenomenon in organic compounds has been known as E-type delayed fluorescence, which was first observed in the eosin dye in 1961 [6]. Figure 8.3 shows examples of organic TADF molecules, including porphyrin derivatives [7], benzophenone [8, 9], aromatic thiones [10], and fullerenes [11, 12], which possess relatively small ΔE_{ST} values.

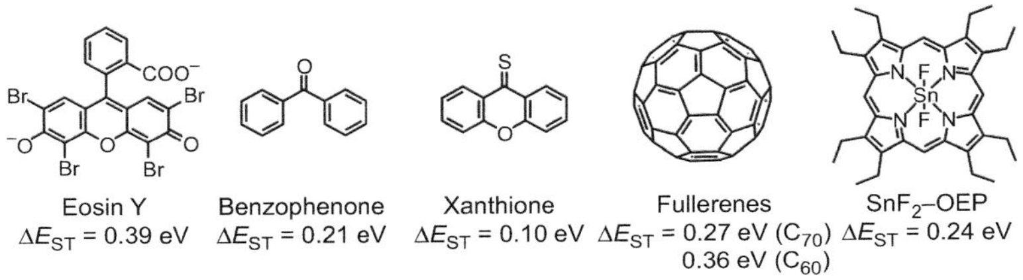

Eosin Y Benzophenone Xanthione Fullerenes SnF$_2$–OEP
ΔE_{ST} = 0.39 eV ΔE_{ST} = 0.21 eV ΔE_{ST} = 0.10 eV ΔE_{ST} = 0.27 eV (C$_{70}$) ΔE_{ST} = 0.24 eV
0.36 eV (C$_{60}$)

Figure 8.3 Classic examples of TADF molecules.

The first application of TADF materials in OLEDs was reported by Adachi et al. in 2009, where **SnF$_2$–OEP** and its derivatives were used as TADF emitters [13]. **SnF$_2$–OEP** shows prompt and delayed fluorescence at around 550–650 nm and weak phosphorescence at around 700 nm at room temperature. As temperature increases, the delayed fluorescence intensity significantly increases, which indicates accelerated RISC by thermal activation. Upon short electrical pulse excitation, both prompt and delayed EL are clearly observed. The delayed component is composed of both TADF and phosphorescence, and the intensity of TADF significantly increases with an increase in temperature. However, the RISC rate constant (k_{RISC}) from T_1 to S_1 is still too small to induce efficient up-conversion, resulting in very low EL efficiencies. Therefore, it is necessary to realize much smaller ΔE_{ST} in order to increase k_{RISC}.

In general, the energies of the S_1 state (E_S) and T_1 state (E_T), and ΔE_{ST} are given by Eqs. (8.3)–(8.5) [5]:

$$E_S = E + K + J \tag{8.3}$$

$$E_T = E + K - J \tag{8.4}$$

$$\Delta E_{ST} = E_S - E_T = 2J > 0 \tag{8.5}$$

where E is the orbital energy, K is the electron repulsion energy, and J is the electron exchange energy. Because J must be positive, E_T is smaller than E_S. In the S_1 or T_1 state, the two unpaired electrons are mainly distributed on the frontier orbitals such as the highest occupied molecular orbital (HOMO) and the lowest unoccupied molecular orbital (LUMO). Therefore, J, which is associated with the Pauli principle, is given by the matrix element (8.6) and can be approximated to be proportional to the overlap integral $\langle \Phi_H | \Phi_L \rangle$ (8.7):

$$J = \iint \Phi_H(r_1) \Phi_L(r_2) \left(\frac{e^2}{r_1 - r_2} \right) \Phi_H(r_2) \Phi_L(r_1) dr_1 dr_2 \quad (8.6)$$

$$J \sim \langle \Phi_H(r_1) \Phi_L(r_2) | \Phi_H(r_2) \Phi_L(r_1) \rangle \sim \langle \Phi_H | \Phi_L \rangle \quad (8.7)$$

where Φ_H and Φ_L are the wave functions of HOMO and LUMO, respectively; e is the charge on an electron; and r_{12} is the distance separating the electrons. According to these equations, the spatial overlap between HOMO and LUMO should be reduced for minimizing ΔE_{ST} and also enhancing RISC.

On the other hand, a relatively large radiative decay rate constant (k_r) from the S_1 state to the S_0 state is also imperative for achieving high Φ_{PL}. However, a large k_r requires a substantial orbital overlap. Since the small ΔE_{ST} and large k_r conflict with each other, the spatial overlap between HOMO and LUMO should be carefully modulated in the design of TADF materials.

8.3 Molecular Design for Efficient TADF Molecules

8.3.1 Intramolecular Donor–Acceptor Systems

The basic design strategy for minimizing ΔE_{ST} is employing an intramolecular donor–acceptor (D–A) system, where its HOMO and LUMO are mainly localized on the electron donor and acceptor moieties, respectively, affording intramolecular charge transfer (CT) excited state. The D–A type TADF molecules can be divided into several classes according to the different connecting manners between donor and acceptor moieties.

The conventional method is that donor and acceptor moieties are directly connected with a highly twisted manner by using steric hindrance and bulkiness. Based on this design, the first designed purely organic TADF molecules, **PIC-TRZ**, were reported by Adachi et al. in 2011 (Fig. 8.4.) [2]. **PIC-TRZ** comprises a 2-biphenyl-1,3,5-triazine acceptor unit coupled with two indolo[2,3-*a*]carbazole donor units. Due to the sterically bulky indolocarbazole substituents, **PIC-TRZ** adopts highly twisted D–A conformation, giving rise to well-separated HOMO and LUMO distributions. Indeed, its ΔE_{ST} value is as small as 0.11 eV. A 6 wt%-doped film of **PIC-TRZ** in 1.3-bis(9-carbazolyl)benzene (**mCP**) as a host shows both prompt and delayed fluorescence with lifetimes of ~10 ns and ~230 μs, respectively, at 300 K. The overall Φ_{PL} is determined to be 39%, which includes the prompt and delayed fluorescence quantum yields of 10% and 29%. OLEDs employing **PIC-TRZ** as the TADF emitter demonstrate a high η_{ext} value of 5.3%, surpassing the theoretical limitation of conventional fluorescent OLEDs. When a fluorescent emitter having Φ_{PL} = 39% is used for OLEDs, the theoretical η_{ext} is estimated to be only 2%. Therefore, it is confirmed that efficient up-conversion of triplet excitons into singlet excitons occurs in the TADF-OLEDs using **PIC-TRZ**. In 2012, Adachi et al. reported a series of groundbreaking efficient TADF molecules based on carbazolyl dicyanobenzene (CDCB), with carbazole as an electron donor and dicyanobenzene as an electron acceptor (Fig. 8.4) [14]. Because the large carbazole units are highly distorted from the dicyanobenzene, the ΔE_{ST} values of these molecules are rather small. Additionally, by changing the electron-donating ability of the peripheral groups, their emission wavelengths can be easily modulated. Indeed, sky-blue to orange TADF emissions were achieved by varying the substitution number and the relative positions of the carbazolyl and cyano groups. For instance, **4CzIPN** shows intense green emission at 507 nm with a high Φ_{PL} of 94% in deoxygenated toluene solution, while **4CzTPN-Ph** and **2CzPN** exhibit orange (577 nm) and sky-blue (473 nm) emissions, respectively. The transient lifetimes of the prompt and delayed fluorescence for **4CzIPN** are 17.8 ns and 5.1 μs, respectively. When oxygen is bubbled through a solution of **4CzIPN**, the lifetime of the delayed fluorescence becomes very short and Φ_{PL} decreases to 10%. These results suggest that the T_1 state is involved in the emission process of **4CzIPN** because the triplet excited state is substantially quenched by triplet oxygen. A 5 wt%-doped film of **4CzIPN** in 4,4'-bis(carbazol-9-yl)biphenyl (**CBP**) shows a high Φ_{PL} of 83% and a very small ΔE_{ST} of 0.083 eV. For the green TADF-OLEDs employing **4CzIPN**, a noticeably high η_{ext} of 19.3% was achieved, which is nearly four times higher than those of conventional fluorescent OLEDs. The orange and sky-blue TADF-OLEDs employing **4CzTPN-Ph** and **2CzPN** also showed high η_{ext} values of 11.2% and 8.0%, respectively. Since this report, a wide variety of combinations of donor and acceptor units with a highly twisted conformation have been exploited to construct efficient D–A type TADF molecules.

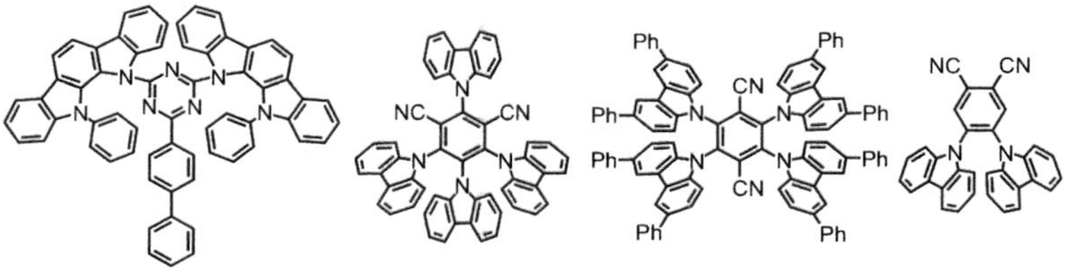

Figure 8.4 Twisted donor–acceptor TADF molecules.

Another method to afford well-separated HOMO and LUMO distributions is that donor and acceptor moieties are connected through non-conjugating linker such as an sp^3-hybridized carbon atom. A spirobifluorene derivative, **Spiro-CN**, is the first TADF molecule involving a spirojunction between donor and acceptor moieties (Fig. 8.5) [15]. **Spiro-CN** has two di-p-tolylamino groups on one fluorene moiety and two cyano groups on the other side. HOMO is localized on the amino-substituted fluorene donor moiety, whereas LUMO is distributed on the cyano-substituted fluorene acceptor moiety. Since the donor and acceptor units are orthogonally connected through the non-conjugating sp^3-hybridized carbon bridge, HOMO and LUMO are spatially well separated from each other. A 6 wt.%-doped film of **Spiro-CN** in mCP shows yellow TADF emission with Φ_{PL} = 27% and ΔE_{ST} = 0.057 eV. In the transient PL measurements, prompt and delayed fluorescence with lifetimes of 24 ns and 14 μs are clearly detected. TADF-OLEDs employing **Spiro-CN** displayed yellow EL and achieved a high maximum η_{ext} of 4.4%. The modified spiro-based TADF molecule, **ACRFLCN**, was reported by the same group in 2012, where acridan was used as a donor unit [16]. A doped film of **ACRFLCN** shows intense TADF with overall Φ_{PL} of 67% and a delayed lifetime of 3.9 ms. The quantum efficiency of

ISC (Φ_{ISC}) is estimated to be 96% for **ACRFLCN**, and the contribution of delayed fluorescence far exceeds that of prompt fluorescence. TADF-OLEDs employing **ACRFLCN** achieved a high maximum η_{ext} of 10.1%. In 2013, a spiro-type TADF molecule, **ACRSA**, possessing an anthracenone acceptor unit was developed [17]. **ACRSA** shows high Φ_{PL} of 81% in the doped film. TADF-OLEDs employing **ACRSA** exhibited blue–green EL with a maximum η_{ext} of 16.5%.

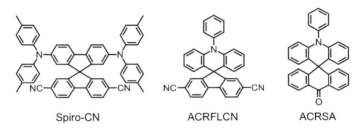

Figure 8.5 Donor–acceptor TADF molecules with spirojunction.

Not only carbon atom but also four-coordinate boron atom can be used as a non-conjugating linker. In 2016, Chou et al. reported a TADF molecule, **PrFTPA**, based on the electron-deficient pyridyl pyrrolide boron complex bearing triphenylamine donor units (Fig. 8.6) [18]. HOMO is mainly located on the triphenylamine moieties, while LUMO is localized on the chelating pyridyl pyrroride moiety. The electron-density contribution of the boron atom in both HOMO and LUMO is almost negligible (0.40% and 0.48%, respectively). This result suggests that the four-coordinate boron atom serves as a node separating its HOMO and LUMO. **PrFTPA** has a small ΔE_{ST} of 0.038 eV and shows TADF emission with Φ_{PL} of 60% in the doped film. TADF-OLEDs employing **PrFTPA** give a maximum η_{ext} of 13.5%. In a follow-up work, the same group synthesized **dfppyBTPA** containing the phenylpyridinato boron complex [19]. While the N,C-chelate phenylpyridinato boron complex is isoelectronic to fluorene, it can act as an electron acceptor because of its higher electron affinity. A doped film of **dfppyBTPA** exhibits highly efficient TADF emission with Φ_{PL} of 100% and the delayed fluorescence lifetime of 2.4 µs. TADF-OLEDs based on **dfppyBTPA** show a considerably high maximum η_{ext} of 26.6%.

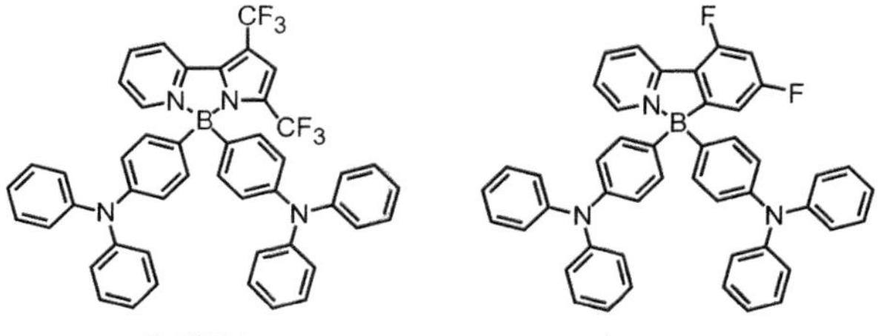

Figure 8.6 Donor–acceptor TADF molecules with four-coordinate boron linker.

In 2015, Swager et al. demonstrated an alternative approach to afford a small ΔE_{ST} utilizing a through-space interaction, wherein π-conjugated systems are in communication by homoconjugation but are sufficiently separated. Based on this strategy, the D–A type triptycenes, **TPA-QNX-(CN)2** and **TPA-PRZ(CN)2**, were designed and synthesized, where diphenylamino groups as donors and dicyanoquinoxaline or dicyanopyrazine moieties as an acceptor were introduced on the different fins of a triptycene scaffold (Fig. 8.7) [20]. Their HOMOs and LUMOs are spatially separated, but their small orbital overlap is simultaneously accomplished by homoconjugation. Small ΔE_{ST} values of 0.111 eV and 0.075 eV are calculated for **TPA-QNX-(CN)2** and **TPA-PRZ(CN)2**, respectively. **TPA-QNX-(CN)2** and **TPA-PRZ(CN)2** exhibit blue–green TADF with Φ_{PL} of 44% and 52%, respectively, in cyclohexane solutions. TADF-OLEDs employing these triptycene emitters display EL with high η_{ext} values of up to 9.4%. As another intramolecular through-space architecture for efficient TADF, Swager et al. designed the D–A molecules **XPT**, **XCT**, and **XtBuCT**, bearing phenothiazine, carbazole or 3,6-di-*tert*-butylcarbazole as a donor unit and 2,4-diphenyl-1,3,5-triazine as an acceptor unit bridged by 9,9-dimethylxanthene (Fig. 8.7) [21]. In these U-shaped molecules, the donor and acceptor units are placed into a cofacial alignment at distances of 3.3–3.5 Å wherein efficient intramolecular through-space CT can occur. HOMO and LUMO are localized on the donor and acceptor moieties, respectively, which gives rise to small ΔE_{ST} of 1–8 meV. TADF characteristics are observed for the xanthene-bridged D–A molecules in both solution and solid states. The TADF-OLEDs employing **XPT** exhibit a maximum η_{ext} of 10%.

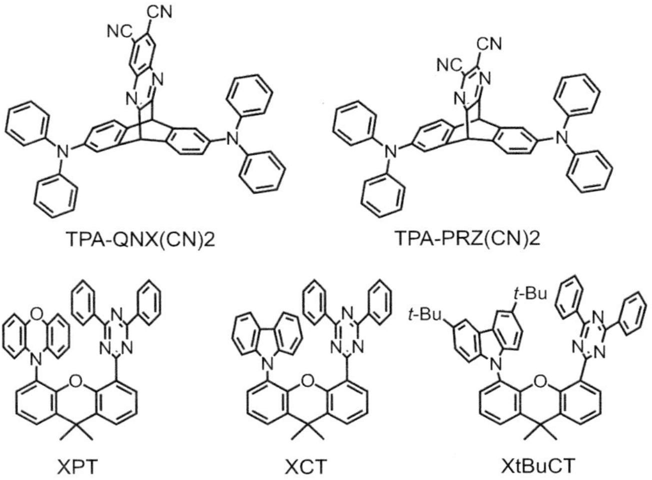

Figure 8.7 Donor–acceptor TADF molecules based on the through-space interactions.

In 2016, Hatakeyama et al. reported a unique design strategy, named multiple resonance effect, for separating HOMO and LUMO and reducing ΔE_{ST} (Fig. 8.8) [22]. The triphenylborane-based molecules **DABNA-1** and **DABNA-2** were synthesized, which have a rigid planar polycyclic aromatic framework with two nitrogen atoms in ortho-substitution from the boron atom. Since the nitrogen atom exhibits the opposite resonance effect to that of the boron atom, HOMO and LUMO can be significantly separated without the need to introduce supplementary donor and acceptor units. LUMO is localized on the boron atom and at the *ortho* and *para* positions relative to it, whereas HOMO is localized on the nitrogen atoms and at the *meta* position relative to the boron atom. Although their ΔE_{ST} values (0.14–0.18 eV) are not so small compared to those of other D–A type TADF molecules, they have a distinct advantage in photophysical properties. Because of the rigid π-conjugated framework with a large oscillator strength for the S_0–S_1 transition of 0.205, **DABNA-1** exhibits a sharp emission spectrum with a much smaller full-width at half-maximum (FWHM) than those of typical D–A type TADF emitters, which results in remarkably high color purity of EL with the CIE coordinates of (0.13, 0.09). TADF-OLEDs employing **DABNA-2** show pure-blue EL centered at 467 nm, with narrow FWHM of 28 nm and high η_{ext} of 20.2%.

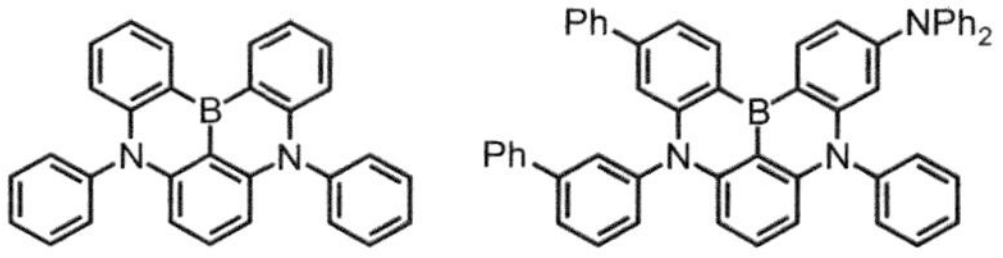

Figure 8.8 Donor–acceptor TADF molecules based on the multiple resonance effect.

8.3.2 Intermolecular Donor–Acceptor Systems (Exciplex TADF)

An exciplex is known as an intermolecular CT state formed between electron-donor and electron-acceptor molecules in the excited state. Since the emission process in an exciplex involves electron transition between the HOMO of the donor and the LUMO of the acceptor, the electron–hole separation distance corresponds to the distance between donor and acceptor molecules. Therefore, the *J* value of the exciplex should become smaller than those intramolecular CT-excited states, leading to a small ΔE_{ST} for facilitating the RISC process.

Adachi et al. first reported an exciplex TADF system by combining 4,4′,4″-tris[3-methylphenyl(phenyl)amino]triphenylamine (***m*-MTDATA**) as a donor and 2-(biphenyl-4-yl)-5-(4-*tert*-butylphenyl)-1,3,4-oxadiazole (***t*-Bu-PBD**) and tris-[3-(3-pyridyl)mesityl]borane (**3TPYMB**) as acceptors (Fig. 8.9) [23]. The PL wavelength of a 50 mol.% ***m*-MTDATA:*t*-Bu-PBD** film is centered at 540 nm, which is much longer than those of the respective neat films of ***m*-MTDATA** and ***t*-Bu-PBD**, indicative of the exciplex formation between these molecules. To realize efficient RISC, the higher triplet energy levels of the donor and acceptor are needed for the exciplex TADF system. Otherwise, the triplet state of the exciplex is easily deactivated via energy transfer to the lower triplet states of the donor or acceptor, followed by non-radiative deactivation. A 50 mol.% ***m*-MTDATA:*t*-Bu-PBD** film exhibits both the prompt and delayed fluorescence. From the temperature dependence of k_{RISC}, the ΔE_{ST} value of the ***m*-MTDATA:*t*-Bu-PBD** exciplex was evaluated to be 50 meV, which implies that the triplet energy level of the exciplex is quite close to their singlet energy level. Exciplex-based OLEDs using the ***m*-MTDATA:*t*-Bu-PBD** system give a maximum η_{ext} of 2%. Since the overall Φ_{PL} is 20% in the 50 mol.% ***m*-MTDATA:*t*-Bu-PBD** film, the theoretical η_{ext} is limited to be approximately 1%, assuming that η_{ST} is 25% and η_{out} is 20–30%. These results indicate that η_{ST} actually increases due to the rapid RISC process in the exciplex system. Moreover, OLED performance can be improved by using a 50 mol.% ***m*-MTDATA:3TPYMB** film. Although the Φ_{PL} is still low (26%), η_{ext} of the **MTDATA:3TPYMB**-based OLED increases to 5.4% because of its high RISC efficiency. To further improve the EL efficiency of exciplex TADF systems, Adachi et al. modified the acceptor to 2,8-bis(diphenylphosphoryl)dibenzo-[*b,d*]thiophene (**PPT**), which has high triplet energy for effective confinement of the triplet exciplex state [24]. OLEDs using 50 mol.% ***m*-MTDATA:PPT** film achieve a high maximum η_{ext} of 10%.

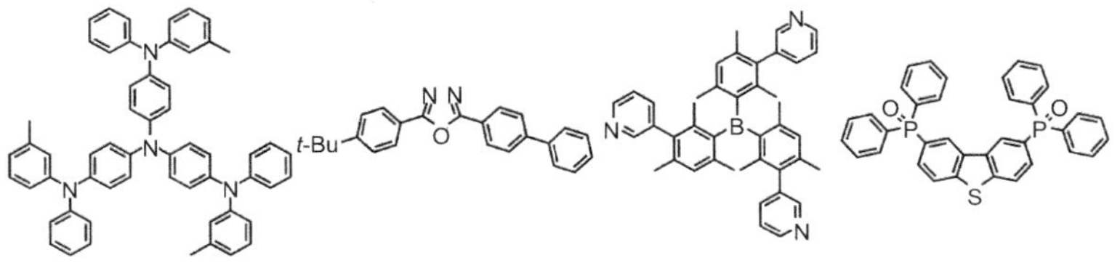

Figure 8.9 Donor and acceptor molecules used in exciplex TADF.

8.4 Design for Functionalized TADF Molecules

8.4.1 TADF Molecules with Aggregation-Induced Emission Property

Organic luminescent materials generally display weakened emission when the molecules aggregate in their condensed solid states, compared to the molecularly isolated state. These aggregation-caused emission quenching is attributed to the enhanced non-radiative decay process by the intermolecular interactions such as exciton coupling, excimer formation, and intermolecular energy transfer. Therefore, to avoid such concentration quenching, most TADF-OLEDs reported so far use doped films as an emission layer, in which TADF molecules are highly dispersed in a suitable host matrix. In this method, elaborate control of the doping concentration in vacuum evaporation is required, which is a drawback in the fabrication of large-area OLED devices for mass production. Thus, the development of TADF molecules exhibiting high PL efficiencies even in their non-doped condensed states is highly desired. One approach is to make TADF molecules have aggregation-induced emission (AIE) characteristics, which is known as an effective approach to suppress concentration quenching and afford efficient solid-state emitters.

Xu et al. reported an asymmetric sulfone-based TADF molecule **SFPC**, which simultaneously exhibits AIE behavior (Fig. 8.10) [25]. While **SFPC** shows very weak PL emission in pure THF solution, the PL intensity increases when the water fraction goes up to 90 vol.% in a THF/water mixture, demonstrating the AIE behavior. The PL emission is enhanced by the formation of the aggregates because of lowering solubility. The Φ_{PL} of **SFPC** in the solid state is estimated to be 93%. **SFPC** also shows strong delayed fluorescence with a lifetime of 1.23 ms in the solid state. Lee et al. reported a similar asymmetric molecule **PTSOPO**, which shows both TADF and AIE characteristics [26]. **PTSOPO** has small ΔE_{ST} of 0.09 eV and exhibits obvious delayed fluorescence with a lifetime of 6.2 μs in the solid state. TADF-OLEDs using a 30 wt.% doped film of **PTSOPO** in bis[2-(diphenylphosphino)phenyl] ether oxide (**DPEPO**) show a maximum η_{ext} of 17.7%, while the non-doped OLEDs using a **PTSOPO** neat film also achieve a high maximum η_{ext} of 17.0%.

Yasuda et al. reported luminescent organic–inorganic conjugated systems based on o-carboranes (1,2-*closo*-dicarbadodecaborane) that possess aggregation-induced delayed fluorescence (AIDF) ability (Fig. 8.10) [27]. o-Carborane is known as an electron-deficient icosahedral boron cluster and an efficient building block for AIE-active systems. The o-carborane derivatives **PCZ-CB-TRZ**, **TPA-CB-TRZ**, and **2PCZ-CB**, bearing electron-donor and electron-acceptor moieties, have been synthesized. In **PCZ-CB-TRZ** and **TPA-CB-TRZ**, HOMOs are located on the respective donor moieties, whereas the corresponding LUMOs are mainly distributed over the peripheral triazine moieties. On the other hand, in the case of **2PCZ-CB**, LUMO is predominantly localized on the central 1,2-diphenylcarborane moiety, which functions as an electron-accepting unit rather than a hub for anchoring the peripheral π-conjugated groups. While these compounds exhibit very weak emissions in THF solution, their neat films show intense emissions with high Φ_{PL} of up to 97% by virtue of their AIE characteristics. Moreover, the transient PL decay of their neat films clearly presents TADF properties. The non-doped OLEDs incorporating these emitters in neat films exhibit a high maximum η_{ext} of up to 11.0%.

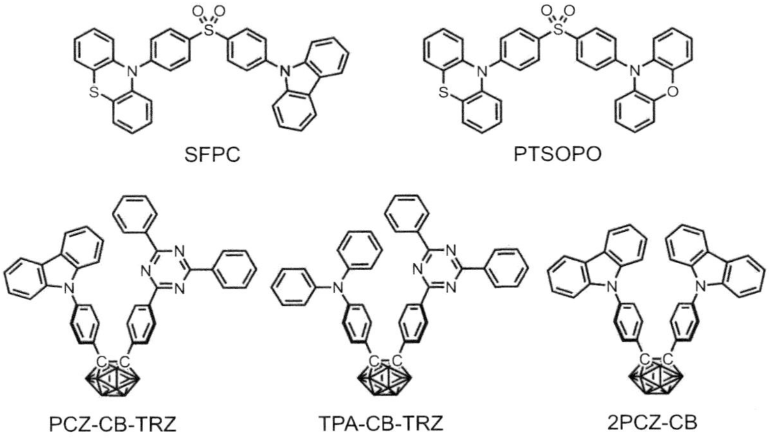

Figure 8.10 AIE-TADF molecules.

8.4.2 TADF Molecules with Stimuli-Responsive Characteristics

Mechanochromic luminescent (MCL) molecules, which exhibit reversible luminescence color changes in response to external stimuli such as mechanical force, temperature, and solvent vapor, have been intensively studied, owing to their potential applications in optoelectronic devices, sensors, probes, and memory devices. By combining MCL and TADF characteristics, multifunctional EL devices can be realized.

Chi et al. reported an asymmetric benzophenone-based TADF molecule **OPC**, which shows both TADF and MCL characteristics

(Fig. 8.11) [28]. **OPC** exhibits a dual emission peaking at 456 and 554 nm in the crystalline solid state, resulting in overall pure white emission with the CIE coordinates of (0.35, 0.35). The transient PL decays reveal that these blue and yellow emissions are normal and delayed fluorescence, respectively. These dual emissions originate from two different conformers; one involves the CT transition between a carbazole moiety and a benzophenone moiety, and the other involves the CT transition between a phenothiazine moiety and a benzophenone moiety. After a solid sample of **OPC** is ground, the PL intensity of blue emission simultaneously declined, leading to clear emission color change from white to yellow. XRD profiles demonstrate that an initial crystalline phase changes into an amorphous state under such mechanical treatment.

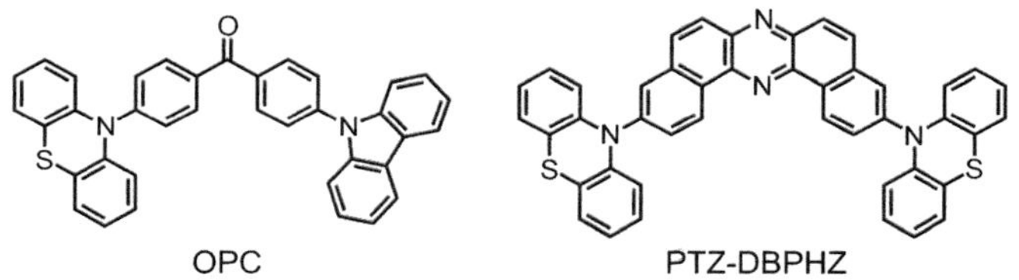

Figure 8.11 MCL-TADF molecules.

Takeda et al. reported a TADF molecule **PTZ-DBPHZ**, which exhibits distinct multi-color changeable MCL characteristics (Fig. 8.11) [29]. **PTZ-DBPHZ** contains dibenzo[*a,j*]phenazine unit as an acceptor coupled with two phenothiazine donor units. Because the phenothiazine unit can exist as two distinct conformers by the conformational alteration, the existence of four possible thermodynamically interconvertible conformers of **PTZ-DBPHZ** can be predicted by the quantum chemical calculations. Indeed, X-ray crystallographic analysis confirms two conformational types of phenothiazines. **PTZ-DBPHZ** clearly shows distinct multi-color-changing MCL behavior in response to a variety of external stimuli. Two polymorphs of **PTZ-DBPHZ** are obtained through recrystallization. One exhibits yellow emission at 568 nm, and the other exhibits orange emission at 640 nm. After these crystals are ground, the emission color changes to deep-red/near IR, with a peak at 673 nm. Thermal annealing of the ground solid yields an orange-emissive solid with an emission wavelength of 646 nm. In contrast, the exposure of CH_2Cl_2 vapor to the ground solid causes a greater blue-shift of the emission wavelength to 596 nm. The interconversion between these states occurs with high reproducibility. This high-contrast MCL behavior results from the switching of the locally excited and intramolecular CT-excited states caused by the conformational change in the phenothiazine units. Moreover, **PTZ-DBPHZ** has a small ΔE_{ST} and exhibits efficient TADF. TADF-OLEDs employing **PTZ-DBPHZ** give a high η_{ext} of 16.8%.

8.4.3 TADF Molecules with Circularly Polarized Luminescence

Circularly polarized luminescence (CPL) is a phenomenon in which optically active compounds emit circularly polarized (CP) light, which is attractive for various applications such as 3D display, quantum computing, spintronics, and asymmetric synthesis. Currently, CP light is often obtained by using CP filter with nonpolarized light source, which leads to a loss in brightness. Thus, OLEDs that can directly emit CP light are desired as an alternative light source. The development of TADF molecules with CPL property can be helpful for enhancing the efficiency of CPL-OLEDs.

Hirata et al. reported the enantiomeric 12-hydroxynaphthacen-5(12H)-one derivative, **DPHN**, which exhibits both CPL and TADF properties (Fig. 8.12) [30]. **DPHN** has a chiral carbon connected with a triphenylamine moiety as a donor and a naphthacenone moiety as an acceptor. HOMO is localized on the donor moiety, while LUMO is localized on the acceptor moiety, and hence the chiral carbon effectively separates HOMO and LUMO. The enantiomer of **DPHN** displays the CPL characteristics with the dissymmetry factor (g_{lum}) of 1.1×10^{-3} in solution, resulting from the chiral configuration between HOMO and LUMO relative to a chiral carbon center. Moreover, **DPHN** shows TADF with Φ_{PL} of 26% in a doped film because of its small ΔE_{ST} of 0.14 eV.

Pieters et al. reported a TADF molecule **BINOL-2CzTPN**, which is capable of emitting CPL through chiral perturbation by a tethered chiral unit (Fig. 8.12) [31]. HOMO is mainly localized on the carbazole moieties, and LUMO is localized on the dicyanobenzene moiety. This orbital distribution implies the non-participation of the

1,1′-binaphtyl moiety in the functioning chromophore, but the close proximity of this chiral unit may efficiently perturb the TADF emitter in order to express optical activity. **BINOL-2CzTPN** shows TADF with Φ_{PL} = 53% and delayed fluorescence lifetime of 2.9 μs in its toluene solution. **BINOL-2CzTPN** also shows CPL characteristics with g_{lum} = 1.3 × 10^{-3}. TADF-OLEDs using **BINOL-2CzTPN** as an emitter show a high η_{ext} of 9.1%. In addition, no racemization has been detected during the fabrication process of the OLED device. Both the high EL efficiency and configurational stability demonstrate that this molecular design is viable for the future development of efficient CPL-OLEDs.

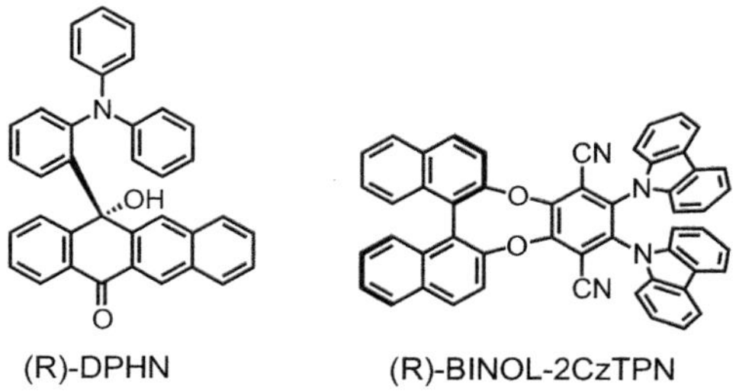

Figure 8.12 CPL-TADF molecules.

8.5 Summary

This chapter describes the molecular design strategies for efficient, purely organic TADF molecules and their PL and EL characteristics. After the first demonstration of TADF-OLEDs, their performance reached the theoretical maximum η_{int} of almost 100% and became comparable to those using organometallic phosphorescent emitters. By taking advantages of the high EL efficiency as well as lower cost, TADF materials have established a solid position as the next-generation luminescent materials. Although many excellent TADF molecules have been developed so far, there are further challenges to improve performance, such as device stability, decline in EL efficiency at high current density, and color purity. We envisioned that the freedom for the TADF molecular design overcomes these problems in the near future.

References

1. Pople, M., Kallmann, H. P., and Magnante, P. (1963). Electroluminescence in organic crystals, *J. Phys. Chem.*, **38**, pp. 2042–2043.
2. Endo, A., Sato, K., Yoshimura, K., Kai, T., Kawada, A., Miyazaki, H., and Adachi, C. (2011). Efficient up-conversion of triplet excitons into a singlet state and its application for organic light emitting diodes, *Appl. Phys. Lett.*, **98**, 083302.
3. Segal, M., Baldo, M. A., Holmes, R. J., Forrest, S. R., and Soos, Z. G. (2003). Excitonic singlet-triplet ratios in molecular and polymeric organic materials, *Phys. Rev. B,* **68**, 075211.
4. Baldo, M., Lamansky, S., Burrows, P., Thompson, M., and Forrest, S. (1999). Nearly 100% internal phosphorescence efficiency in an organic light-emitting devices, *Appl. Phys. Lett.*, **75**, pp. 4–6.
5. Turro, N. J., Ramamurthy, V., and Scaiano, J. C. (1978). *Modern Molecular Photochemistry* (University Science Book, USA).
6. Parker, C. A. and Hatchard, C. G. (1961). Triplet-singlet emission in fluid solutions. Phosphorescence of eosin, *Trans. Faraday Soc.*, **57**, pp. 1894–1904.
7. Parker, C. A. and Joyce, T. A. (1967). Delayed fluorescence and some properties of the chlorophyll triplets, *Photochem. Photobiol.*, **6**, pp. 395–406.
8. Jones, P. F. and Calloway, A. R. (1971). Temperature effects on the intramolecular decay of the lowest triplet state of benzophenone, *Chem. Phys. Lett.,* **10**, pp. 438–443.
9. Wolf, M. W., Legg, K. D., Brown, R. E., Singer, L. A., and Parks, J. H. (1975). Photophysical studies on the benzophenones. Prompt and delayed fluorescences and self-quenching, *J. Am. Chem. Soc.*, **97**, pp. 4490–4497.
10. Maciejewski, A., Szymanski, M., and Steer, R. P. (1986). Thermally activated delayed S_1 fluorescence of aromatic thiones, *J. Phys. Chem.*, **90**, pp. 6314–6318.
11. Berberan-Santos, M. N. and Garcia, J. M. M. (1996). Unusually strong delayed fluorescence of C70, *J. Am. Chem. Soc.*, **118**, pp. 9391–9394.
12. Salazar, F. A., Fedorov, A., and Berberan-Santos, M. N. (1997). A study of thermally activated delayed fluorescence in C60, *Chem. Phys. Lett.*, **271**, pp. 361–366.
13. Endo, A., Ogasawara, M., Takahashi, A., Yokoyama, D., Kato, Y., and Adachi, C. (2009). Thermally activated delayed fluorescence from Sn^{4+}–porphyrin complexes and their application to organic light-

emitting diodes: A novel mechanism for electroluminescence, *Adv. Mater.*, **21**, pp. 4802–4806.

14. Uoyama, H., Goushi, K., Shizu, K., Nomura, H., and Adachi, C. (2012). Highly efficient organic light-emitting diodes from delayed fluorescence, *Nature*, **492**, pp. 234–238.

15. Nakagawa, T., Ku, S.-Y., Wong, K.-T., and Adachi, C. (2012). Electroluminescence based on thermally activated delayed fluorescence generated by a spirobifluorene donor–acceptor structure, *Chem. Commun.*, **48**, pp. 9580–9582.

16. Méhes, G., Nomura, H., Zhang, Q., Nakagawa, T., and Adachi, C. (2012). Enhanced electroluminescence efficiency in a spiro-acridine derivative through thermally activated delayed fluorescence, *Angew. Chem. Int. Ed.*, **51**, pp. 11311–11315.

17. Nasu, K., Naagawa, T., Nomura, H., Lin, C.-J., Cheng, C.-H., Tseng, M.-R., Yasuda, T., and Adachi, C. (2013). A highly luminescent spiro-anthracenone-based organic light-emitting diode exhibiting thermally activated delayed fluorescence, *Chem. Commun.*, **49**, pp. 10385–10387.

18. Shiu, Y.-J., Cheng, Y.-C., Tsai, W.-L., Wu, C.-C., Chao, C.-T., Lu, C.-W., Chi, Y., Chen, Y.-T., Liu, S.-H., and Chou, P.-T. (2016). Pyridyl pyrrolide boron complexes: The facile generation of thermally activated delayed fluorescence and preparation of organic light-emitting diodes, *Angew. Chem. Int. Ed.*, **55**, pp. 3017–3021.

19. Shiu, Y.-J., Chen, Y.-T., Lee, W.-K., Wu, C.-C., Lin, C.-T., Liu, S.-H., Chou, P.-T., Lu, C.-W., Cheng, I.-C., Lien, Y.-J., and Chi, Y. (2017). Efficient thermally activated delayed fluorescence of functional phenylpyridinato boron complexes and high performance organic light-emitting diodes, *J. Mater. Chem. C*, **5**, pp. 1452–1462.

20. Kawasumi, K., Wu, T., Zhu, T., Chae, H. S., Voothis, T. V., Baldo, M. A., and Swager, T. M. (2015). Thermally activated delayed fluorescence materials based on homoconjugation effect of donor–acceptor triptycenes, *J. Am. Chem. Soc.*, **137**, pp. 11908–11911.

21. Tsujimoto, H., Ha, D.-G., Markopoulos, G., Chae, H. S., Baldo, M. A., and Swager, T. M. (2017). Thermally activated delayed fluorescence and aggregation induced emission with through-space charge transfer, *J. Am. Chem. Soc.*, **139**, pp. 4894–4900.

22. Hatakeyama, T., Shiren, K., Nakajima, K., Nomura, S., Nakatsuka, S., Kinoshita, K., Ni, J., Ono, Y., and Ikuta, T. (2016). Ultrapure blue thermally activated delayed fluorescence molecules: Efficient HOMO–LUMO separation by the multiple resonance effect, *Adv. Mater.*, **28**, pp. 2777–2781.

23. Gousi, K., Yoshida, K., Sato, K., and Adachi, C. (2012). Organic light-emitting diodes employing efficient reverse intersystem crossing for triplet-to-singlet state conversion, *Nat. Photon.*, **6**, pp. 253–258.
24. Gousi, K. and Adachi, C. (2012). Efficient organic light-emitting diodes through up-conversion from triplet to singlet excited states of exciplexes, *Appl. Phys. Lett.*, **101**, 023306.
25. Xu, S., Liu, T., Mu, Y., Wang, Y.-F., Chi, Z., Lo, C.-C., Liu, S., Zhang, Y., Lien, A., and Xu, J. (2015). An organic molecule with asymmetric structure exhibiting aggregation-induced emission, delayed fluorescence, and mechanoluminescence, *Angew. Chem. Int. Ed.*, **54**, pp. 874–878.
26. Lee, I. H., Song, W., and Lee, J. Y. (2016). Aggregation-induced emission type thermally activated delayed fluorescent materials for high efficiency in non-doped organic light-emitting diodes, *Organic Electronics*, **29**, pp. 22–26.
27. Furue, R., Nishimoto, T., Park, I. S., Lee, J., and Yasuda, T. (2016). Aggregation-induced delayed fluorescence based on donor/acceptor-tethered janus carborane triads: Unique photophysical properties of nondoped OLEDs, *Angew. Chem. Int. Ed.*, **55**, pp. 7171–7175.
28. Xie, Z., Chen, C., Xu, S., Li, J., Zhang, Y., Liu, S., Xu, J., and Chi, Z. (2015). White-light emission strategy of a single organic compound with aggregation-induced emission and delayed fluorescence properties, *Angew. Chem. Int. Ed.*, **54**, pp. 7181–7184.
29. Okazaki, M., Takeda, Y., Data, P., Pander, P., Higginbotham, H., Monkman, A. P., and Minakata, S. (2017). Thermally activated delayed fluorescent phenothiazine-dibenzo[a,j]phenazine-phenothiazine triads exhibiting tricolor-changing mechanochromic luminescence, *Chem. Sci.*, **8**, pp. 2677–2686.
30. Imagawa, T., Hirata, S., Totani, K., Watanabe, T., and Vacha, M. (2015). Thermally activated delayed fluorescence with circularly polarized luminescence characteristics, *Chem. Commun.*, **51**, pp. 13268–13271.
31. Feuillastre, S., Pauton, M., Gao, L., Desmarchelier, A., Riives, A. J., Prim, D., Tondelier, D., Geffroy, B., Muller, G., Clavier, G., and Pieters, G. (2016). Design and synthesis of new circularly polarized thermally activated delayed fluorescence emitters, *J. Am. Chem. Soc.*, **138**, pp. 3990–3993.

Chapter 9

α-Diketone-Type Precursors of Acenes for Solution-Processed Organic Solar Cells

Mitsuharu Suzuki and Hiroko Yamada

Division of Materials Science, Graduate School of Science and Technology, Nara Institute of Science and Technology 8916-5 Takayama-cho, Ikoma, Nara 630-0192, Japan
hyamada@ms.naist.jp

9.1 Introduction

Acenes, the aromatic hydrocarbons comprising linearly fused benzenes (Fig. 9.1a), have long been serving as benchmark compounds in the development of organic semiconductors [1]. The superiority of acenes lies in their rigid, planar π-frameworks that enable effective intermolecular contact and efficient charge-carrier transport in the solid state. For example, good field-effect hole mobilities of ~1 cm^2/Vs have been routinely obtained in pentacene (**1**, Fig. 9.1b), which are comparable to typical mobilities observed in amorphous silicon [2]. At the same time, however, the structural characteristics of acenes bring about problems in their synthesis, purification, and processing. Specifically, the chemical stability and solubility drastically decrease along with an increase in the molecular size. Pentacene is already hardly soluble in any organic solvents and easily oxidized in ambient conditions, prohibiting its processing by conventional solution techniques. Accordingly, fabrication of pentacene-based semiconductor devices largely depends on costly vacuum processes. In the cases of pristine acenes larger than pentacene, even isolation is elusive in the literature [3, 4].

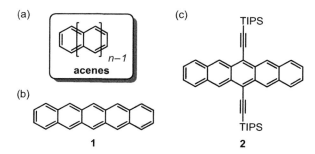

Figure 9.1 (a) General structure of acenes. (b, c) Chemical structure of pentacene (**1**) and the soluble pentacene derivative **2**. TIPS: triisopropylsilyl.

A straightforward solution to this problem would be the introduction of substituents that can confer stability and solubility. The pentacene derivative having two triisopropylsilylethynyl substituents at the 6,13-positions (**2**, Fig. 9.1c) is a prime example in this context [5]. This relatively simple derivative holds a stability and solubility high enough for solution processing in ambient conditions. When this compound was processed into highly crystalline thin films via a carefully optimized solution-shearing method, an excellent hole mobility of 11 cm^2/Vs was achieved [6]. Many other soluble derivatives of pentacene, including oligomers and polymers, followed the prosperous development of compound **2**, although successful examples are rather limited in terms of achieving high charge-carrier mobility [7, 8]. A potential problem associated with this approach is that solubilizing substituents can attenuate effective π–π contacts in the solid state and unpredictably change the packing motif or film morphology.

An alternative way to evade the stability and solubility problems is the 'precursor approach', wherein a precursor compound is processed into a thin film and then quantitatively converted in situ to a target material by external nonchemical stimuli (i.e., heat or light). If the precursor possesses adequate stability and solubility, one can easily purify and deposit it by well-established solution-based techniques. While several different types of such precursors of acenes are known [9], this chapter focuses on α-diketone-type photoprecursors [10]. The following sections cover the basic characteristics of this class of precursors and their use in the preparation of organic solar cells comprising acenes as active-layer components.

9.2 Synthesis of Acenes from α-Diketone-Type Photoprecursors

The photochemical conversion of α-diketone-type precursors to corresponding acene compounds is typically achieved upon the n–π* excitation of the α-diketone moiety (Fig. 9.2a). The n–π* excitation can be induced with visible light around 470 nm, and thus this reaction is very mild, requiring neither high-energy light nor harsh thermal treatment. Furthermore, this reaction is essentially "traceless" because the only byproduct is gaseous carbon monoxide, which can be easily removed from the reaction system either in solution or in the solid state. These characteristics are highly desirable in preparing semiconducting thin films for electronic applications.

Although the first reports of this photoinduced decarbonylative aromatization were published as early as the 1960s [12, 13], it was not until 2005 that this reaction was applied to pentacene (Fig. 9.2b) [11, 14]. Soon after, the scope of this chemistry was successfully extended to larger analogues up to nonacene (**3**) by the groups of Neckers and Bettinger (Fig. 9.2c) [15–17]. Note that the generation of **3** from its α-diketone-type precursor **3-DK** is a special case in which higher energy light of <320 nm is required and the reaction is performed at a very low temperature of 30 K in an argon matrix [17].

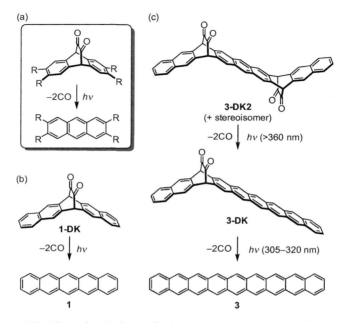

Figure 9.2 Photochemical synthesis of acenes from α-diketone-type precursors. (a) General scheme. (b) Generation of pentacene (**1**) from its precursor **1-DK** [11]. (c) Stepwise decarbonylation of bis(α-diketone) **3-DK2** to nonacene (**3**) [17].

9.3 Preparation of Acene Thin Films by in situ Conversion of α-Diketone-Type Photoprecursors

The α-diketone-type photoprecursors are generally more soluble than the corresponding pristine acenes. In addition, their photochemical conversion to acenes proceeds efficiently even in the thin-film state. These characteristics make it possible to prepare thin films of insoluble acenes via an indirect solution process in which α-diketone-type precursors are wet processed into thin films and then converted to corresponding acene compounds by photo-irradiation.

While this *photoprecursor approach* requires additional costs of synthesizing precursor compounds and carrying out the in situ photoreaction, it holds multiple advantages over direct solution deposition. Apparently, the applicability of solution-based processing techniques is largely expanded because of the enhanced solubility of α-diketone derivatives. The higher solubility can be ascribed primarily to the prevention of extensive intermolecular interactions upon the structural change from planar acenes to bent α-diketone derivatives. Another benefit of the photoprecursor approach is that the structural parameters of thin films can be tuned depending on photoreaction conditions. It may be even possible to obtain thin films of high quality that cannot be accessed by conventional solution processes. These aspects are illustrated below with specific examples.

9.3.1 Photoprecursor Approach for Solution Processing of Insoluble Acenes

The photoprecursor approach can afford high-quality thin films of insoluble acenes that can be employed as active layers in organic devices. This is well exemplified by organic field-effect transistors (OFETs) based on pentacene (**1**) [18]. OFETs of the bottom-gate-top-contact configuration can be fabricated via the photoprecursor approach as follows (Fig. 9.3): (i) Spin-coating of the precursor **1-DK** on a pretreated Si/SiO$_2$ substrate; (ii) irradiation with a blue LED (λ = 470 ± 10 nm) to effect the decarbonylative aromatization of **1-DK** to form **1**; (iii) thermal evaporation of gold as source and drain electrodes. The resulting OFETs showed the maximum field-effect hole mobility of 0.86 cm^2/Vs. This value is as high as those typically obtained for vacuum-deposited pentacene films (~1 cm^2/Vs) [2]. In addition, it is one of the best results among those obtained via precursor approaches and direct solution deposition using rather harsh conditions (hot trichlorobenzene as cast solvent) (Table 9.1) [19–22].

The mobility of 0.86 cm^2/Vs was achieved with carefully optimized deposition conditions. First, a mixed solvent system containing both low- and high-boiling-point solvents was employed in spin coating for achieving a fine balance between the crystallinity and thickness of the resulting films. Screening of different solvent systems found that chloroform with 1% trichlorobenzene gave an optimal result. Second, the intensity of light for the photoreaction from **1-DK** to **1** was tuned such that the photoreaction could complete before the spin-coated film dried up. Third, the substrate was warmed during the photoreaction for controlling the degree of crystallization. The optimized photoreaction conditions were light intensity of 300 mW/cm^2, irradiation time of 60 min, and substrate temperature of 80°C.

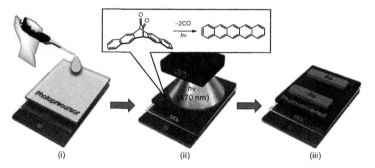

Figure 9.3 Schematic description of the fabrication process of pentacene-based FETs through the photoprecursor approach. (i) Solution deposition of **1-DK**, (ii) photoreaction of **1-DK** to **1**, (iii) thermal evaporation of gold electrodes to complete the device. Adapted with permission from Ref. [10], Copyright 2013, Elsevier.

Table 9.1 Field-effect hole mobilities in pentacene deposited by precursor approaches

Precursor	Hole mobility (cm^2/Vs)	$I_{on/off}$	Conversion conditions	Reference
1-DK	0.86	4.3 × 10^6	470 nm at 80°C	18
4	0.89	>2 × 10^7	200°C	19
5	0.25	>8 × 10^4	UV, then 130°C	20
6	0.38	~10^6	210°C	21
—*	0.45	>10^5	—	22

*Pentacene was deposited as a solution in hot trichlorobenzene.

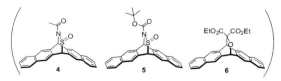

It should also be noted that the best hole mobility was observed in a relatively low-crystallinity film. This result may be explained by the attenuated negative effect of the domain boundary. Namely, the pentacene film prepared under the optimized conditions contains both amorphous and crystalline regions, and boundaries between crystalline domains are presumably filled with amorphous pentacene that can still transport charge carriers. As a result, more carriers may be able to reach the electrodes without being fully trapped at boundaries between crystallites.

9.3.2 Photoprecursor Approach as an Alternative Means for Processing Soluble Acenes

The comparative examination of differently prepared thin films of tetrahexylanthratetrathiophene **7** (Fig. 9.4a) further elucidated the characteristics of the photoprecursor approach [23]. Compound **7** is well soluble owing to the four *n*-hexyl chains, and thus directly solution processable. It is also possible to introduce the α-diketone unit onto the central anthracene framework and process compound **7** through the photoprecursor approach. In addition, **7** is thermally stable so that it can be processed by thermal evaporation under vacuum to form thin films. Comparison of thus-obtained thin films revealed that the photoprecursor approach could afford a higher-quality film than vacuum deposition. Specifically, the surface roughness probed by atomic force microscopy (AFM) was smaller for the film prepared by the photoprecursor approach than the vacuum-deposited film (the root mean square (RMS) roughness values were 0.7 and 1.9 nm, Table 9.2). Furthermore, the hole mobility determined by the space-charge-limited current (SCLC) method was higher for the former (2.1×10^{-5} cm^2/Vs) than the latter (1.8×10^{-5} cm^2/Vs). On the other hand, the film prepared through direct solution deposition was quite rough, having large grains and boundaries (RMS roughness = 8.5 nm). As a result, the SCLC hole mobility in this film could not be determined because the effective film thickness was hard to define.

Table 9.2 Surface roughness* of thin films of compounds **7** and **8** [23]

Deposition method	7	8
Photoprecursor approach	0.7 nm	19.2 nm
Vacuum deposition	1.9 nm	1.7 nm
Direct solution deposition	8.5 nm	16.4 nm

*Determined as the RMS value of surface roughness observed by tapping-mode AMF.

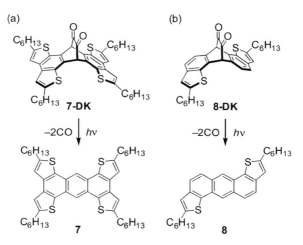

Figure 9.4 Chemical structures of anthratetrathiophene **7**, anthradithiophene **8**, and their α-diketone-type photoprecursors [23].

The high smoothness of the film prepared by the photoprecursor approach can be linked to the nature of the precursor, which cannot pack effectively during deposition because of the bent structure. In other words, molecules of **7-DK** loosely pack in the thin film without excessive formation of aggregates or crystallites, thereby affording a highly smooth surface. Upon the photoconversion, the molecular conformation changes from bent to planar, but the morphology of the resulting thin film largely maintains the smoothness of the precursor film because of the limited mobility of molecules in the solid state. In contrast, the highly self-aggregating nature of compound **7** is strongly reflected in its directly solution-processed thin film, most likely because molecules are quite mobile during deposition, with the aid of cast solvent.

On the other hand, the photoprecursor approach does not lead to a better morphology as compared to direct solution deposition in the case of bent anthradithiophene derivative **8** (Table 9.2). The RMS roughness of a directly solution-deposited film of derivative **8** is 16.4 nm, while it is 19.2 nm for the counterpart prepared by the photoprecursor approach. For reference, the RMS roughness is one order lower for a vacuum-deposited film (1.66 nm). The contrasting results between compounds **7** and **8** can be attributed to the difference in molecular size. A smaller molecule has higher mobility during the photoreaction in thin films, leading to the formation of larger aggregates and rough surface. This example well demonstrates the effectiveness and limitations of the photoprecursor approach over direct solution deposition.

9.4 Preparation of Organic Photovoltaic Layers via the Photoprecursor Approach

The photoprecursor approach can be employed for the preparation of not only single-component thin films of acenes but also blend films, as far as the partner compound does not completely inhibit the photoinduced decarbonylative aromatization. For example, acene-based semiconductors can be co-deposited with fullerene derivatives through this approach. Many acene compounds are known to work as p-type semiconductor, while fullerene derivatives such as [6,6]-phenyl-C_{61}-butyric acid methyl ester ($PC_{61}BM$) and [6,6]-phenyl-C_{71}-butyric acid methyl ester ($PC_{71}BM$) are among the most studied organic n-type semiconductors. As the active layer of organic solar cells is typically a p–n intermixed (or bulk heterojunction, BHJ) thin film, a properly selected acene–fullerene combination may form an excellent photovoltaic active layer. This section provides two examples in this framework of concept.

9.4.1 Preparation of BHJ Layers

The first example concerns the blend of 2,6-di(2-thienyl)anthracene (DTA) and $PC_{71}BM$ (Fig. 9.5) [24]. DTA can be processed into thin films through the photoprecursor approach using the corresponding α-diketone-type derivative DTADK as a soluble precursor. Organic solar cells (OSCs) comprising the DTA:PC$_{71}$BM blend can be prepared similarly to the case of OFETs as follows: (i) Spin coating of a solution of DTA:PC$_{71}$BM blend on a pretreated glass/indium tin oxide (ITO) substrate, (ii) irradiation with a blue LED to effect the photoreaction of DTADK to DTA, (iii) thermal evaporation of a cathode electrode metal through a shadow mask.

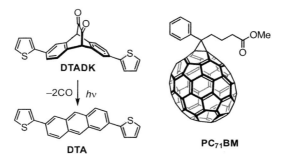

Figure 9.5 Chemical structures of p-type material DTA, its photoprecursor DTADK, and n-type material PC$_{71}$BM.

The photoprecursor DTADK is not much self-aggregating and well miscible with PC$_{71}$BM, thereby affording homogeneous, smooth thin films when spin coated from a chloroform solution. On the other hand, DTA is strongly self-aggregating because of its high planarity and rigidity. Accordingly, DTA:PC$_{71}$BM blend films, after the photoreaction, are typically associated with excessive phase separation, which leads to poor photovoltaic performance. This is, of course, not limited to DTA, but quite general to planar, rigid, and small-sized acenes that lack flexible alkyl substituents. A question here is whether it is possible to improve such a far-from-ideal morphology to gain respectable photovoltaic performance by taking advantage of the photoprecursor approach.

It is well known that post-deposition annealing is highly effective and often crucial for obtaining high-performance BHJ photovoltaic layers comprising a well-soluble p-type material and a fullerene-based n-type material [25]. Highly solubilizing alkyl groups on solution-processable p-type materials prevent excessive self-aggregation and ensure a high degree of miscibility with n-type materials. Thus, the post-deposition annealing for a majority of solution-processed BHJ layers is intended to induce an adequate degree of phase separation. In contrast, DTA is not well miscible with PC$_{71}$BM and forms large grains by self-aggregation. Thus, the morphology of DTA:PC$_{71}$BM is improved by inducing phase mixing rather than phase separation. Indeed, the commonly employed morphology-improvement techniques were not found particularly effective for the DTA:PC$_{71}$BM blend; namely, the thermal annealing resulted in moderate improvements at the best, while the solvent-vapor annealing and the use of a small amount of high-boiling-point additive (1,8-diiodooctane) led to a deterioration or very limited improvement in the photovoltaic performance [24].

Relevant morphological improvement in DTA:PC$_{71}$BM was achieved when *ortho*-dichlorobenzene (ODCB) was added to the cast solution in chloroform as a co-solvent. As summarized in Table 9.3, the short-circuit current density (J_{SC}) and fill factor (FF) considerably improved along with the increase in the amount of ODCB co-solvent from 0% to 10% and to 20%. Accordingly, the power-conversion efficiency (PCE) increased from 0.44% to 1.79% and to 2.18%. Improvements in J_{SC} and FF are usually attributed to the enhancement in charge-carrier generation and/or charge-carrier transport efficiencies, which are strongly affected by the active-layer morphology.

Table 9.3 Photovoltaic parameters of BHJ OPVs based on the DTA:PC$_{71}$BM blend prepared via the photoprecursor approach*

Entry	ODCB (vol.%)	Active-layer thickness (nm)	J_{SC} (mA/cm^2)	V_{OC} (V)	FF (%)	PCE (%)
1	0	106	1.75	1.03	24.4	0.44
2	10	90	4.62	0.97	39.9	1.79
3	20	66	4.90	0.93	47.7	2.18

*Parameters of the best-performed cell in each entry.
General device structure: [ITO/PEDOT:PSS/DTA:PC$_{71}$BM (2:1)/Ca/Al].
Source: Ref. [24].

The open-circuit voltage (V_{OC}) was, in contrast to J_{SC} and FF, relatively unaffected by the addition of ODCB co-solvent. Nonetheless, slight differences are noticeable; namely, the V_{OC} gradually decreased from 1.03 V to 0.93 V as the amount of ODCB increased from 0% to

20%. There have been several reports in which V_{OC} slightly decreases upon film annealing associated with significant increase in J_{SC} and FF [26–28]. In particular, Palomares et al. postulated in their report that the difference in crystallinity affected the charge recombination dynamics and the electronic structure of the p-type material, and thus V_{OC} [28]. This claim may be valid also to the DTA:PC$_{71}$BM system.

The crystallinity and molecular orientation in thin films can be analyzed through two-dimensional grazing-incident wide-angle X-ray diffraction (2D GIWAXD) analysis (Fig. 9.6). The film deposited with 20% ODCB showed a series of strong diffractions around q_z = 0.33, 0.65, and 1.0 Å^{-1} (Fig. 9.6c). These diffractions can be assigned to the 200, 400, and 600 diffractions of DTA [29], indicating the end-on stacking of DTA molecules. There are also relatively strong rings around q = 1.3 (overlapping with the diffraction of PC$_{71}$BM), 1.6, and 1.9 Å^{-1}, which correspond well with the 111, 020, and 121 diffractions of DTA, respectively [29]. As the b- and c-axes correspond to the DTA herringbone stack, the observed rings indicate that this staking motif is randomly oriented in the film. Such a structure is more favorable than a selective edge-on or end-on packing in terms of achieving out-of-plane charge-carrier transport. On the other hand, the broad halos around q = 1.3 Å^{-1}, which most likely originated from crystallites of PC$_{71}$BM [30], indicated that the crystallinity of PC$_{71}$BM domains did not change significantly by the use of the co-solvent. Note that the intensities of these halos are hard to quantify because of the overlapping diffraction from DTA crystallites as described above, and thus this discussion remains rather qualitative.

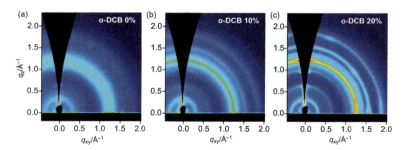

Figure 9.6 GIWAXD images of the DTA:PC$_{71}$BM blend films deposited by the photoprecursor approach on glass/ITO/PEDOT:PSS. The amounts of ODCB co-solvent in cast solutions are (a) 0%, (b) 10%, and (c) 20%. Adapted with permission from Ref. [24], Copyright 2016, American Chemical Society.

In contrast to the clear difference in crystallinity, the surface roughness changed minimally upon the addition of ODCB co-solvent. The surface morphology probed with AFM revealed that the RMS roughness values were 9.3, 10.6, and 10.6 nm for the films prepared with 0, 10, and 20% ODCB, respectively (Fig. 9.7). At the same time, however, apparent differences did exist among these films: Grains aggregated randomly when no ODCB was used, while they assembled to form flower-like objects when 20% ODCB was added. The film prepared using 10% ODCB showed "in-between" characteristics of the other two.

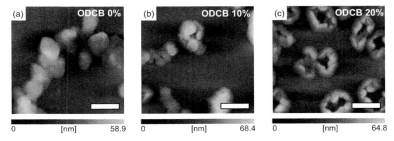

Figure 9.7 Tapping-mode AFM images of BHJ films deposited on glass/ITO/PEDOT:PSS by the photoprecursor approach with different amounts of ODCB co-solvent. (a) 0%, RMS roughness = 9.3 nm; (b) 10%, RMS roughness = 10.6 nm; (c) 20%, RMS roughness = 10.6 nm. The scale bars correspond to 0.5 µm. Adapted with permission from Ref. [24], Copyright 2016, American Chemical Society.

On the other hand, the films before photoreaction (i.e., DTADK:PC$_{71}$BM blend films) all showed highly smooth surface (Fig. 9.8). This observation clearly demonstrates that the photoprecursor DTADK is not as self-aggregating as DTA and well miscible with PC$_{71}$BM under the employed deposition conditions. In addition, the comparison of pre- and post-photoreaction images indicates that molecules can move considerably during the photoreaction in the solid state, and the mobility of molecule seems depending on the amount of ODCB co-solvent. Note that the films spin coated with ODCB co-solvent were immediately subjected to the photoreaction when they were still wet. It seems reasonable to assume that the enhanced dynamics of molecules during the photoreaction in wet films has caused the significant change in crystallinity, morphology, and thus photovoltaic performance of the resulting blend films. It is also worth noting that a neat film of DTA prepared by the photoprecursor approach using 20% ODCB co-solvent showed completely different morphology as compared to the corresponding DTA:PC$_{71}$BM blend film. Thus, not only the residual ODCB but also PC$_{71}$BM should have played an important role in the formation of the observed morphology.

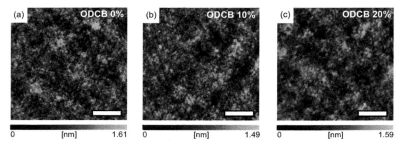

Figure 9.8 Tapping-mode AFM images of the DTADK:PC$_{71}$BM films spin coated on glass/ITO/PEDOT:PSS using different amounts of ODCB co-solvent. (a) 0%, RMS roughness = 0.25 nm; (b) 10%, RMS roughness = 0.23 nm; (c) 20%, RMS roughness = 0.29 nm. The scale bars correspond to 0.5 μm. Adapted with permission from Ref. [24], Copyright 2016, American Chemical Society.

Further investigation of the thin films by fluorescence microspectroscopy revealed that the use of ODCB co-solvent led to the effective suppression of the radiative quenching of excitons in DTA domains, presumably because of the decrease in domain size and increase in p–n interface, both of which facilitate the conversion of excitons to hole–electron pairs.

Overall, this example demonstrated that the morphology in DTA:PC$_{71}$BM blend was largely improved by performing the post-deposition photoreaction when the film was still wet. The resulting films showed enhanced J_{SC} and FF, and thus PCE, due to the increase in crystallinity and the decrease in domain size. As these two structural changes are usually not compatible (i.e., increase in crystallinity is often associated with increase in domain size), the photoprecursor approach provides unique opportunities in morphology control of organic small-molecule blend films.

9.4.2 Application to Layer-by-Layer Solution Deposition

The second example is regarding the preparation of "p–i–n" photovoltaic active layers. In this type of active layers, a p:n blend (intermixed layer or i-layer) is sandwiched between neat p- and n-type semiconductors (p- and n-layers) (Fig. 9.9). This triple-layer structure is expected to possess the strengths of both the p–n bilayer and BHJ monolayer structures: namely, efficient generation of charge carriers in the i-layer and swift extraction of the charge carriers through the p- and n-layers. The superiority of p–i–n structure was computationally predicted by Ray et al. [31], and experimentally pursued by, among others, the groups of Hiramoto and Reo by employing vacuum-deposited active layers with or without dopants [32, 33].

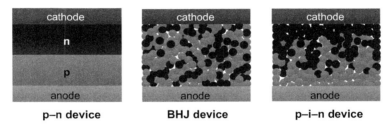

Figure 9.9 Schematic drawing of the three types of photovoltaic active layers mentioned in the text. The dark and light colors in the active layers represent n-type and p-type materials, respectively.

However, the p–i–n structure is hard to prepare with conventional solution processes because of dissolution of the underlying layer upon deposition of another layer. Although several methods such as the use of orthogonal solvents or post-deposition crosslinking have been proposed in order to evade the dissolution problem, each method, more or less, has limitations in the scope and applicability. Thus, there is still much room to explore new methodologies.

The photoprecursor approach is highly promising in this context because it can bring about a high contrast in solubility before and after the photoinduced decarbonylation, as described in Sec. 9.3.1. Specifically, if the solubility of the photoreaction product is low enough, one can deposit another material on top of it as a solution.

Yamaguchi et al. indeed examined the applicability of photoprecursor approach by employing DTA and PC$_{71}$BM as active-layer components [34].

The deposition process of p–i–n photovoltaic layers via the photoprecursor approach is schematically illustrated in Fig. 9.10, along with those of BHJ and p–n layers. The p-layer was prepared by spin coating of DTADK in chloroform followed by photo-irradiation to effect the in situ conversion of DTADK to DTA. Since DTA is essentially insoluble in chloroform, it is possible to solution deposit another layer on it using the same solvent. Thus, the i-layer was then prepared in the same manner as the p-layer but using a DTADK:PC$_{71}$BM blend solution instead of DTADK. As shown earlier in Sec. 9.4.1, the photoconversion of DTADK to DTA proceeds even in the existence of PC$_{71}$BM. Now, the resulting i-layer is still partly insoluble, enabling the deposition of PC$_{71}$BM in chloroform as the n-layer on top it. Note that one can tune the thickness of each sublayer by changing the concentration of cast solution.

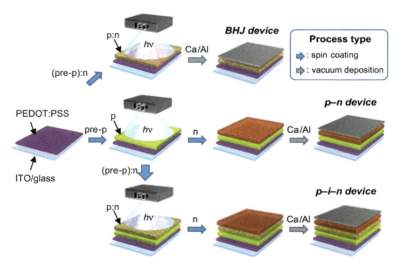

Figure 9.10 Schematic illustration of preparation of BHJ, p–n, p–i–n OSCs through the photoprecursor approach. Adapted from Ref. [34].

Table 9.4 summarizes the photovoltaic performance of thus-obtained p–n, BHJ, and p–i–n devices. The superiority of the p–i–n device over the other two is apparent by comparing entries 1, 2, and 4. The p–n device showed a good FF of 53.6%, while that of the BHJ device is only 29.3%, reflecting the high efficiency of the p–n structure in charge-carrier extraction. On the other hand, the BHJ device showed a slightly higher J_{SC} of 2.92 mA/cm^2 than that of the p–n device (2.81 mA/cm^2) despite the inferior charge-carrier extraction. This is most likely reflecting the more efficient charge-carrier generation in the BHJ active layer. The p–i–n device showed, as expected, both high J_{SC} and FF to afford the best PCE among these three devices. Interestingly, the PCE in the p–i–n device was improved (1.38 → 1.50%) when the thickness of the i-layer was increased by using a cast solution of a higher concentration (10 → 20 mg/mL). This is in contrast to the case of BHJ system, for which PCE decreased (0.90 → 0.47%) upon an increase in the active-layer thickness because of the high resistance in the BHJ layer with unfavorable morphology (see Sec. 9.4.1).

Table 9.4 Photovoltaic parameters of three different types of OSCs comprising DTA and PC$_{71}$BM as p-type and n-type materials, respectively*

Entry	Device type	Solution concentration	Active-layer thickness (nm)	J_{SC} (mA/cm^2)	V_{OC} (V)	FF (%)	PCE (%)
1	p–n	5/10	75	2.81	0.80	53.6	1.22
2	BHJ	10	67	2.92	1.05	29.3	0.90
3		20	124	1.82	1.01	25.4	0.47
4	p–i–n	5/10/10	123	3.64	0.82	46.1	1.38
5		5/20/10	153	3.78	0.89	44.7	1.50

*Parameters of the best-performed cell in each entry.
General device structure: [ITO/PEDOT:PSS/active layer/Ca/Al].
Source: Ref. [34].

All these observations have clearly demonstrated the effectiveness of the p–i–n structure in enhancing the photovoltaic performance of otherwise poorly performing BHJ active layers. This result also manifests the effectiveness of the photoprecursor approach in the controlled preparation of organic photovoltaic active layers via a layer-by-layer solution process. Considering the structural simplicity of DTA, it can be expected that elaboration of the compound may easily lead to a considerable improvement in photovoltaic efficiency. If one employs PCBM as the n-type material, the primary considerations in designing p-type materials for p–i–n OSCs are as follows: (i) The p-layer material should be high in transparency and hole-transport capability, (ii) the p-type material for the i-layer should have high photoabsorption capability and an adequate degree of miscibility with PCBM, (iii) the highest occupied molecular orbital energy levels should match between p-type materials in the p- and i-layers for ensuring barrier-less extraction of holes. These considerations strongly suggest that using different p-type materials between the p- and i-layers is desirable for achieving an optimal p–i–n system. Similarly, it would be better to use different n-type materials between the i- and n-layers. The validity of this strategy has been shown in a preliminary study [34].

9.5 Summary and Outlook

This chapter described the photochemical synthesis of acenes from corresponding α-diketone-type precursors and their application in the preparation of acene-based semiconducting thin films. The resulting films can be high enough in quality to serve as active layers in organic electronic devices such as OFETs and OSCs. A question to be answered here would be why we should even care this approach if it causes additional costs in chemical synthesis (preparation of photoprecursors) and in deposition (irradiation to effect photoreaction). Following are the (potential) benefits that, we assume, can bring more than compensation of the additional costs:

- It enables us to solution process insoluble acene compounds, which are, otherwise, difficult to isolate in the pure form.
- It is very mild and requires only visible-light irradiation for the post-deposition in situ photoreaction.
- It can be applicable to a variety of acene compounds because the α-diketone unit can be installed, in principle, if the compound has a benzene ring, which is the most common structural motif in organic semiconductors.
- It frees molecules from extensive decoration with flexible solubilizing substituents and allows them to have more efficient intermolecular π–π contacts and thus higher charge-

carrier transport capability. This can also lead to higher morphological stability than conventional systems.
- It enables effective control over the vertical composition profile in solution-processed multi-component thin films.
- It provides new possibilities in morphology control through the tuning of photoreaction conditions.

Some of these were already demonstrated as mentioned in this chapter, and some others are left to be proved in future work. If we can take advantage of all these benefits, a significant advance in the development of organic electronics will be realized.

References

1. Anthony, J. E. (2008) The larger acenes: Versatile organic semiconductors, *Angew. Chem. Int. Ed.*, **47**(3), pp. 452–483.
2. Kitamura, M. and Arakawa, Y. (2008) Pentacene-based organic field-effect transistors, *J. Phys. Condens. Matter*, **20**(18), 184011.
3. Tönshoff, C. and Bettinger, H. F. (2013) Beyond pentacenes: Synthesis and properties of higher acenes. In: *Polyarenes I Topics in Current Chemistry* (Springer, Berlin, Heidelberg), pp. 1–30.
4. Dorel, R. and Echavarren, A. M. (2017) Strategies for the synthesis of higher acenes, *Eur. J. Org. Chem.*, **2017**(1), pp. 14–24.
5. Anthony, J. E., Brooks, J. S., Eaton, D. L., and Parkin, S. R. (2001) Functionalized pentacene: Improved electronic properties from control of solid-state order, *J. Am. Chem. Soc.*, **123**(38), pp. 9482–9483.
6. Diao, Y., Tee, B. C.-K., Giri, G., Xu, J., Kim, D. H., Becerril, H. A., Stoltenberg, R. M., Lee, T. H., Xue, G., Mannsfeld, S. C. B., and Bao, Z. (2013) Solution coating of large-area organic semiconductor thin films with aligned single-crystalline domains, *Nat. Mater.*, **12**(7), pp. 665–671.
7. Lehnherr, D. and Tykwinski, R. R. (2010) Oligomers and polymers based on pentacene building blocks, *Materials*, **3**(4), pp. 2772–2800.
8. Anthony, J. E. (2006) Functionalized acenes and heteroacenes for organic electronics, *Chem. Rev.*, **106**(12), pp. 5028–5048.
9. Watanabe, M., Chen, K.-Y., Chang, Y. J., and Chow, T. J. (2013) Acenes generated from precursors and their semiconducting properties, *Acc. Chem. Res.*, **46**(7), pp. 1606–1615.
10. Suzuki, M., Aotake, T., Yamaguchi, Y., Noguchi, N., Nakano, H., Nakayama, K., and Yamada, H. (2014) Synthesis and photoreactivity of α-diketone-

type precursors of acenes and their use in organic-device fabrication, *J. Photochem. Photobiol. C Photochem. Rev.*, **18**, pp. 50–70.

11. Uno, H., Yamashita, Y., Kikuchi, M., Watanabe, H., Yamada, H., Okujima, T., Ogawa, T., and Ono, N. (2005) Photo precursor for pentacene, *Tetrahedron Lett.*, **46**(12), pp. 1981–1983.

12. Strating, J., Zwanenburg, B., Wagenaar, A., and Udding, A. C. (1969) Evidence for the expulsion of bis-CO from bridged α-diketones, *Tetrahedron Lett.*, **10**(3), pp. 125–128.

13. Bryce-Smith, D. and Gilbert, A. (1968) Photochemical and thermal cycloadditions of cis-stilbene and tolan (diphenylacetylene) to tetrachloro-p-benzoquinone. Photodecarboxylation of an α-diketone, *Chem. Commun. Lond.*, **0**(21), pp. 1319–1320.

14. Yamada, H., Yamashita, Y., Kikuchi, M., Watanabe, H., Okujima, T., Uno, H., Ogawa, T., Ohara, K., and Ono, N. (2005) Photochemical synthesis of pentacene and its derivatives, *Chem. Eur. J.*, **11**(21), pp. 6212–6220.

15. Mondal, R., Adhikari, R. M., Shah, B. K., and Neckers, D. C. (2007) Revisiting the stability of hexacenes, *Org. Lett.*, **9**(13), pp. 2505–2508.

16. Mondal, R., Tönshoff, C., Khon, D., Neckers, D. C., and Bettinger, H. F. (2009) Synthesis, stability, and photochemistry of pentacene, hexacene, and heptacene: A matrix isolation study, *J. Am. Chem. Soc.*, **131**(40), pp. 14281–14289.

17. Tönshoff, C. and Bettinger, H. F. (2010) Photogeneration of octacene and nonacene, *Angew. Chem. Int. Ed.*, **49**(24), pp. 4125–4128.

18. Nakayama, K., Ohashi, C., Oikawa, Y., Motoyama, T., and Yamada, H. (2013) Characterization and field-effect transistor performance of printed pentacene films prepared by photoconversion of a soluble precursor, *J. Mater. Chem. C*, **1**(39), pp. 6244–6251.

19. Afzali, A., Dimitrakopoulos, C. D., and Breen, T. L. (2002) High-performance, solution-processed organic thin film transistors from a novel pentacene precursor, *J. Am. Chem. Soc.*, **124**(30), pp. 8812–8813.

20. Weidkamp, K. P., Afzali, A., Tromp, R. M., and Hamers, R. J. (2004) A photopatternable pentacene precursor for use in organic thin-film transistors, *J. Am. Chem. Soc.*, **126**(40), pp. 12740–12741.

21. Chao, T.-H., Chang, M.-J., Watanabe, M., Luo, M.-H., Chang, Y. J., Fang, T.-C., Chen, K.-Y., and Chow, T. J. (2012) Solution processed high performance pentacene thin-film transistors, *Chem. Commun.*, **48**(49), pp. 6148–6150.

22. Minakata, T. and Natsume, Y. (2005) Direct formation of pentacene thin films by solution process, *Synth. Met.*, **153**(1), pp. 1–4.
23. Quinton, C., Suzuki, M., Kaneshige, Y., Tatenaka, Y., Katagiri, C., Yamaguchi, Y., Kuzuhara, D., Aratani, N., Nakayama, K., and Yamada, H. (2015) Evaluation of semiconducting molecular thin films solution-processed via the photoprecursor approach: The case of hexyl-substituted thienoanthracenes, *J. Mater. Chem. C*, **3**(23), pp. 5995–6005.
24. Suzuki, M., Yamaguchi, Y., Takahashi, K., Takahira, K., Koganezawa, T., Masuo, S., Nakayama, K., and Yamada, H. (2016) Photoprecursor approach enables preparation of well-performing bulk-heterojunction layers comprising a highly aggregating molecular semiconductor, *ACS Appl. Mater. Interfaces*, **8**(13), pp. 8644–8651.
25. Liu, Y., Zhao, J., Li, Z., Mu, C., Ma, W., Hu, H., Jiang, K., Lin, H., Ade, H., and Yan, H. (2014) Aggregation and morphology control enables multiple cases of high-efficiency polymer solar cells, *Nat. Commun.*, **5**, 5293.
26. Kniepert, J., Lange, I., Heidbrink, J., Kurpiers, J., Brenner, T. J. K., Koster, L. J. A., and Neher, D. (2015) Effect of solvent additive on generation, recombination, and extraction in PTB7:PCBM solar cells: A conclusive experimental and numerical simulation study, *J. Phys. Chem. C*, **119**(15), pp. 8310–8320.
27. Collins, B. A., Li, Z., Tumbleston, J. R., Gann, E., McNeill, C. R., and Ade, H. (2013) Absolute measurement of domain composition and nanoscale size distribution explains performance in PTB7:PC71BM solar cells, *Adv. Energy Mater.*, **3**(1), pp. 65–74.
28. Sánchez-Díaz, A., Izquierdo, M., Filippone, S., Martin, N., and Palomares, E. (2010) The origin of the high voltage in DPM12/P3HT organic solar cells, *Adv. Funct. Mater.*, **20**(16), pp. 2695–2700.
29. Meng, H., Sun, F., Goldfinger, M. B., Jaycox, G. D., Li, Z., Marshall, W. J., and Blackman, G. S. (2005) High-performance, stable organic thin-film field-effect transistors based on bis-5′-alkylthiophen-2′-yl-2,6-anthracene semiconductors, *J. Am. Chem. Soc.*, **127**(8), pp. 2406–2407.
30. Lu, L., Chen, W., Xu, T., and Yu, L. (2015) High-performance ternary blend polymer solar cells involving both energy transfer and hole relay processes, *Nat. Commun.*, **6**, 7327.
31. Ray, B. and Alam, M. A. (2012) Random vs regularized OPV: Limits of performance gain of organic bulk heterojunction solar cells by morphology engineering, *Sol. Energy Mater. Sol. Cells*, **99**, pp. 204–212.

32. Hiramoto, M., Suemori, K., Matsumura, Y., Miyata, T., and Yokoyama, M. (2006) P-I-N junction organic solar cells, *Mol. Cryst. Liq. Cryst.*, **455**(1), pp. 267–275.
33. Siebert-Henze, E., Lyssenko, V. G., Fischer, J., Tietze, M., Brueckner, R., Schwarze, M., Vandewal, K., Ray, D., Riede, M., and Leo, K. (2014) Built-in voltage of organic bulk heterojunction p-i-n solar cells measured by electroabsorption spectroscopy, *AIP Adv.*, **4**(4), 047134.
34. Yamaguchi, Y., Suzuki, M., Motoyama, T., Sugii, S., Katagiri, C., Takahira, K., Ikeda, S., Yamada, H., and Nakayama, K. (2014) Photoprecursor approach as an effective means for preparing multilayer organic semiconducting thin films by solution processes, *Sci. Rep.*, **4**, 7151.

Chapter 10

Photochemical Synthesis of Phenacenes and Their Application to Organic Semiconductors

Hideki Okamoto[a] and Yoshihiro Kubozono[b]

[a]*Division of Earth, Life, and Molecular Sciences, Graduate School of Natural Science and Technology, Okayama University, Okayama 700-8530, Japan*
[b]*Research Institute for Interdisciplinary Science, Okayama University, Okayama 700-8530, Japan*
hokamoto@okayama-u.ac.jp

10.1 Introduction

10.1.1 General Remarks on Significance of Polycyclic Aromatic Hydrocarbons (PAHs) in Material Sciences

The importance of aromatic molecules with extended π-conjugation has increased owing to their fundamental chemical and physical natures as well as their potential applications in various functional materials such as organic field-effect transistors (OFETs) [1–4],

---

Light-Active Functional Organic Materials
Edited by Hiroko Yamada and Shiki Yagai
Copyright © 2019 Jenny Stanford Publishing Pte. Ltd.
ISBN 978-981-4800-15-0 (Hardcover), 978-0-429-44853-9 (eBook)
www.jennystanford.com

organic light-emitting diodes (OLEDs) [5, 6], and organic photovoltaics (OPVs) [7–9]. By using organic molecules, flexible, light-weight, and large-area electronic devices can be fabricated and the device fabrication costs and energies can be suppressed. Since the discovery that a pentacene-based OFET device showed very high hole mobility (μ = 1.5 cm^2/Vs) [10], organic electronics with polycyclic aromatic molecules drastically accelerated. When we look for literatures in SciFinder with the keywords OFET, OPV, and OLED, we find that thousands of publications appear year after year (Fig. 10.1), which recognize that organic compounds play a significant role in future electronic materials. Additionally, it is worth mentioning that some polycyclic aromatic compounds, such as picene, phenanthrene, and dibenzopentacene, provide new aromatic-compound-based superconductors [11–13]. Therefore, polycyclic aromatic molecules possess great potentials to open novel material sciences. Acenes, typified by pentacene, have been extensively investigated in the fields of synthetic and theoretical chemistry as well as material sciences in the last two decades. In contrast, much less effort has been paid to the chemistry, physics, and material sciences of phenacenes. (See Fig. 10.2 for general chemical structures of phenacenes.) However, phenacenes have been revealed to serve as novel materials in organic electronics. In this chapter, we describe phenacene-based material chemistry, including physical properties, synthesis, and semiconductor applications by comparing with those of acenes.

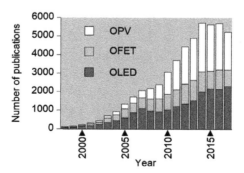

Figure 10.1 Increasing publications dealing with organic electronic materials.

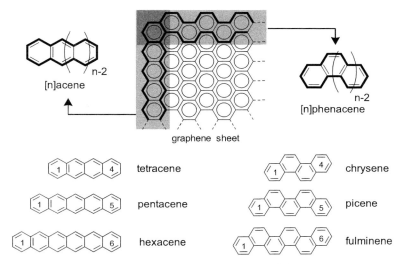

Figure 10.2 General chemical structures of acene and phenacene.

10.1.2 Chemical Structures of PAHs: Acenes and Phenacenes

First, let us confirm the chemical structures of acenes and phenacenes. As for the chemical structures of PAHs, there are many structural categories classified by the shapes of the molecules. Two of the major structural series of PAHs are depicted in Fig. 10.2. One is *acene*s in which benzene rings are fused in a linear array. The other is *phenacene*s in which benzene rings are aligned in an alternative W-shaped array. Acenes and phenacenes of the same benzene rings are isomers of each other. The acene structure is considered to be a ribbon fragment of the zigzag graphene edge, while the phenacene structure corresponds to the ribbon fragment of the arm-chair edge of graphene.

For the acene and phenacene series, the structures for lower homologues (n = 4–6) are displayed with their names. Larger phenacenes ($n \geq 7$) are called [n]phenacenes as they have no trivial name.

10.2 Physical Properties of Phenacenes

10.2.1 Stability of Acene and Phenacene

As generally recognized, large acenes ($n \geq 5$) are unstable under ambient conditions. In contrast, phenacenes are quite stable and robust in an atmosphere of UV illumination in the presence of oxygen under which acenes are never durable. The difference in the relative stabilities between the two series can be qualitatively understood by their aromatic natures [14]. Figure 10.3 shows Kekulé (upper) and Clar (lower) descriptions of pentacene and picene [15]. The benzene rings possessing six-electron aromatic sextet are drawn with a six-membered ring with a circle inside. The arrow in Clar's pentacene structure indicates that only one of the five benzene rings has the aromatic sextet. In contrast, picene has three rings with aromatic sextet and two isolated double bonds. Thus, phenacenes are considered to be more stable than the corresponding acenes due to the aromatic stability. Theoretical calculations have predicted that larger acenes have singlet open-shell character in the ground state, but the open-shell electronic features of acenes are still controversial [16].

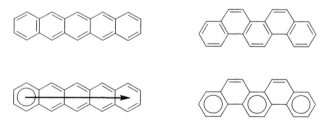

Figure 10.3 Kekulé (upper) and Clar (lower) structures of pentacene and picene.

10.2.2 Electronic Features of Acenes and Phenacenes

Figure 10.4 shows the electronic absorption spectra of anthracene and phenanthrene. These spectra display differences in optical behavior between acene and phenacene series [17, 18]. For anthracene, the longer-wavelength absorption band (340–380 nm) is assigned to the ^1L_a band possessing molecular absorption coefficient $\varepsilon \sim 1 \times 10^4$ M^{-1}cm^{-1}. In contrast, phenanthrene displays the first absorption band in the longer-wavelength region (320–340 nm) with much smaller ε value ($\varepsilon \sim 10^2$ M^{-1}cm^{-1}) and the second band at around 300 nm ($\varepsilon \sim 10^4$ M^{-1}cm^{-1}). The former and the latter absorption bands are, respectively, assigned to ^1L_b and ^1L_a electronic transitions. Similar absorption behavior is also observed for larger phenacenes such as picene and fulminene; thus, phenacenes display very weak ^1L_b absorption band in their longest absorption wavelength region, as shown in Fig. 10.5a [19, 20]. The ^1L_a band (p band by Clar's expression) is considered to correspond to the highest occupied molecular orbital (HOMO)–lowest unoccupied molecular orbital (LUMO) electronic transition. Optical band gaps of acenes and phenacenes estimated from the energy of the p band are shown in Fig. 10.5b as a function of number of benzene rings [n] [19–21]. It is obvious that the p band absorption wavelength for acenes drastically red-shifts with an increase in the benzene ring number, whereas that for phenacenes only slightly red-shifts depending on [n]. These facts indicate that phenacenes keep high optical band gap even if the phenacene π-conjugation extends. The experimentally determined HOMO–LUMO energy gaps of large phenacenes (n = 5–9) and pentacene are shown in Fig. 10.6 [22, 23]. The HOMO energy levels of phenacenes are lower than those of pentacene, reflecting their chemical stability. The HOMO–LUMO gap of phenacenes decreases only slightly with an increase in the benzene ring numbers ranging from 3.3 eV for picene to 2.9 eV for [9]phenacene.

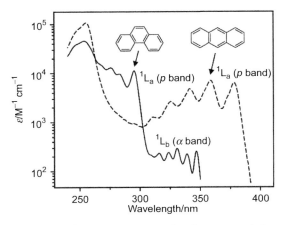

Figure 10.4 Electronic absorption spectra of anthracene and phenanthrene.

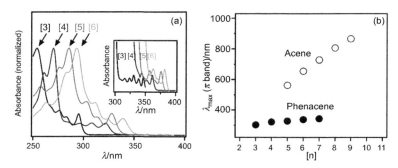

Figure 10.5 (a) Electronic absorption spectra of [n]phenacenes in CHCl$_3$. The inset illustrates the small-intensity absorption bands in the longer-wavelength region. (b) Dependence of absorption (p band) wavelengths on the benzene ring numbers for acene (○) and phenacenes (●). The absorption data for phenacenes are taken from Refs. [19] and [20], and those for acenes are taken from Ref. [21] and the literatures therein.

10.2.3 Solid-State Structure of Phenacenes

The crystal structure of picene is illustrated in Fig. 10.7 [24]. Picene molecules align in a herringbone packing to form a two-dimensional layer within the *ab* plane (Fig. 10.7a). The long axis of picene molecule is aligned parallel to the *b* axis (Fig. 10.7b). The picene crystal packing is similar to that reported for pentacene [25, 26]. The X-ray diffraction patterns for higher phenacenes have been observed by using their powder and/or thin-film samples, and the crystallographic parameters have been collected [11, 23, 27, 28]. Table 10.1 summarizes the lattice parameters. For the series of higher phenacenes, the *c* axis length elongates with an increase in the benzene ring numbers, and the *a* and *b* axes lengths are similar for the phenacene series investigated. Thus, the long axis of the phenacene molecules aligns along with the *c* axis of the unit cell as in the case of picene (Fig. 10.7). The *c* axis distance linearly correlates with the benzene ring numbers of the phenacene molecules, as shown in Fig. 10.8. In the case of fulminene, the *a* and *b* lattice constants are larger than those of other phenacenes, suggesting that the arrangement of the molecule in the *ab* plane is different from the case of the other phenacene molecules.

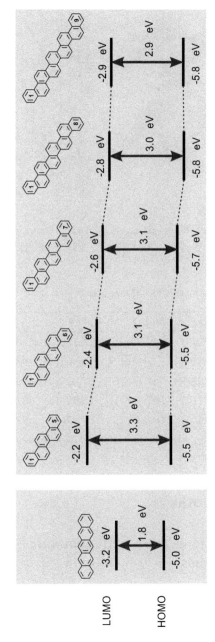

Figure 10.6 HOMO–LUMO energy levels and the energy gaps of pentacene and [n]phenacenes (n = 5–9) [22, 23].

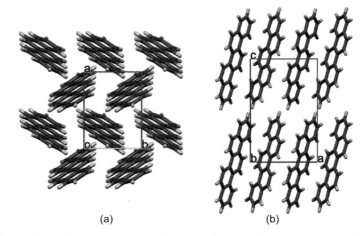

Figure 10.7 Crystal structures of picene: (a) two-dimensional herringbone packing in the *ab* plane, (b) crystal packing viewed along with the *b* axis. Reproduced with the CIF data in Ref. [24].

Table 10.1 Lattice parameters for [*n*]phenacenes

	a (Å)	*b* (Å)	*c* (Å)	β (°)	Ref.
Picene[a]	8.472(2)	6.170(2)	13.538(7)	90.81(4)	[11]
Fulminene[a]	12.130(1)	7.9416(7)	15.401(1)	93.161(8)	[27]
[7]phenacene	8.4381(8)	6.1766(6)	17.829(2)	93.19(1)	[27]
[8]phenacene	8.842(2)	6.043(1)	19.896(4)	92.92(3)	[28]
[9]phenacene	8.844(5)	6.127(3)	22.47(1)	92.72(5)	[23]

[a]Picene and fulminene, respectively, refer to [5]- and [6]phenacenes.

10.3 Photochemical Synthesis of Phenacene Frameworks

10.3.1 Conventional Systematic Synthetic Methods of Phenacenes

Higher phenacenes, such as picene and fulminene, were identified as stable aromatic compounds in the oil industry more than hundred years ago [29, 30]. They are, thus, very old compounds. However, as they had not been used as functional materials for a long time, efficient and systematic synthesis of phenacenes was rarely investigated until the discovery that picene served as the active layer of high-performance OFET devices [31]. Figure 10.9 summarizes some conventional systematic phenacene syntheses found in the literature. Harvey reported that acid-catalyzed Friedel–Crafts type intramolecular cyclization of the precursor A and B followed by dehydrogenetion using DDQ (2,3-dichloro-5,6-dicyano-1,4-benzoquinone) or Pd/C effectively produced picene and fulminene [32]. Photoreaction of diarylethanes C and D produced the corresponding phenacene skeletons by photoreaction in the presence of 9-fluorenone as a photosensitizer in low chemical yields [20, 33]. The most versatile synthetic reaction is considered to be the photocyclization of diarylethenes E followed by oxidative aromatization of the intermediary generated dihydrophenacenes F. Iodine and molecular oxygen are conventionally used as the oxidant to efficiently afford the phenacene products. Currently, new oxidation strategies alternative to the I_2/O_2 system are being developed to effectively promote the photochemical synthesis of phenacenes [34]. This type of reaction is referred to as the Mallory photocyclization [35]. Figure 10.10 shows the general reaction scheme for this photocyclization. Hundreds of examples of Mallory photocyclization are found in previously published review article [36], and one can apply this reaction to construct various PAH skeletons. Additionally, this synthetic strategy has an advantage: The diarylethene precursors can be easily obtained by Wittig reaction between aryl aldehydes (Ar-CHO) and (arylmethy)triphenylphosphonium salts (Ar-CH$_2$PPh$_3$X).

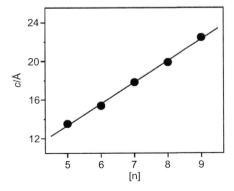

Figure 10.8 Plots of the c axis length as a function of benzene ring numbers [n] of phenacene (cf. Table 10.1).

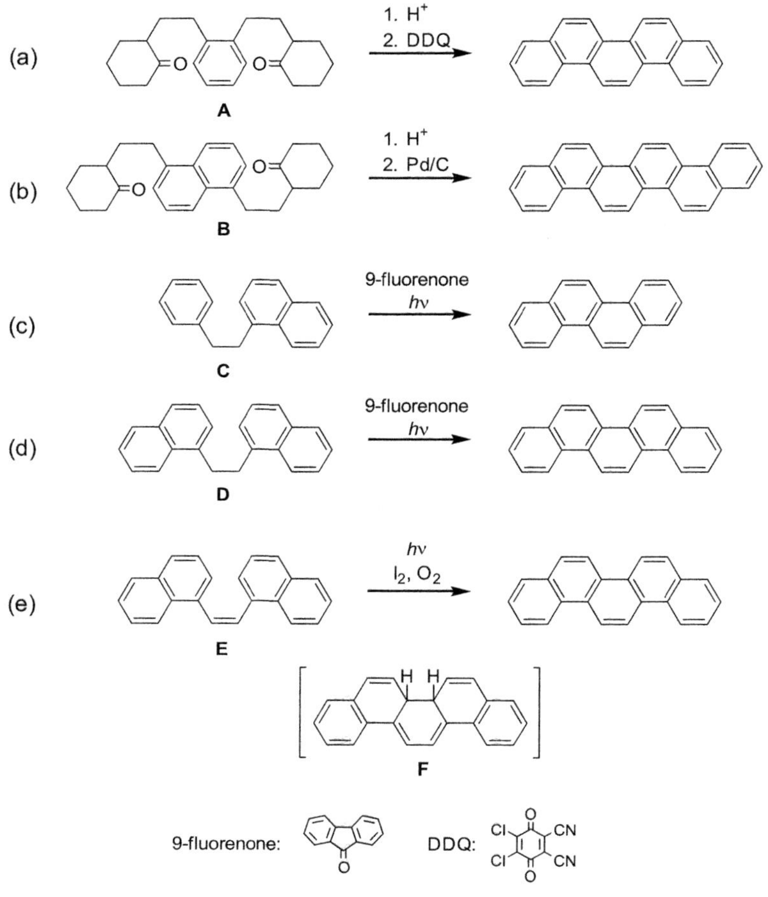

Figure 10.9 Conventional systematic synthesis of phenacene frameworks.

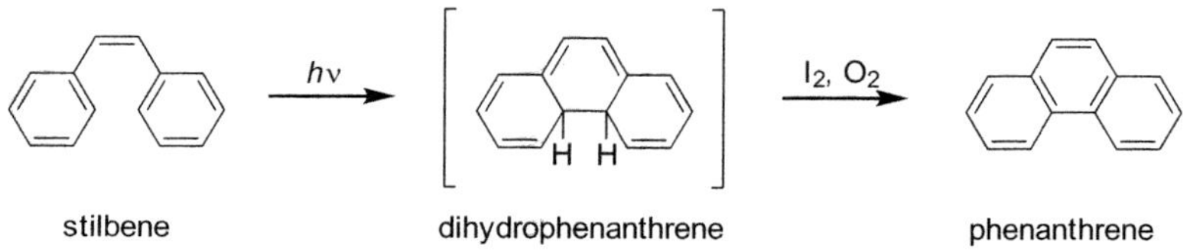

Figure 10.10 General reaction scheme of Mallory photocyclization.

10.3.2 Synthesis of Large Phenacenes by Mallory Photocyclization

By using the Mallory photocyclization, large phenacenes can be prepared with ease, as shown in Fig. 10.11. The photocyclizing part of the diarylethene precursors are highlighted with thick bond lines. Mallory has synthesized [7]phenacene in good yield from bis(1-phenanthryl)ethene (Fig. 10.11, left). Although the solubility of [7]phenacene in common solvents was very poor, it was successfully characterized by spectroscopic methods [19, 37, 40]. Later, [8]- and [9]phenacenes were also successfully synthesized, as shown in Fig. 10.11 (center and right) [23, 28]. Because these phenacenes are completely insoluble in common organic solvents, they precipitated from the reaction solution during the photolysis. As they could not be spectroscopically characterized in the solution phase, their structures were assigned by solid-state analytical methods, such as X-ray diffraction (Table 10.1), MALDI TOF mass spectrometry, and elemental analysis. Specifically, the molecular shape of [9]phenacene was observed by scanning tunneling microscopy (STM) on Ag(111) surface to provide direct evidence for its molecular structure [39]. Figure 10.12 shows the STM images of [7]- and [9]phenacenes. Currently, [9]phenacene is the largest phenacene possessing no substituent.

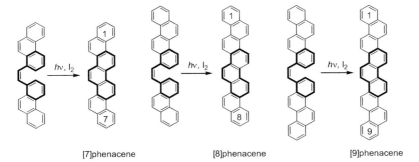

Figure 10.11 Photochemical synthesis of large phenacenes.

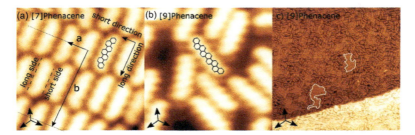

Figure 10.12 STM images of [7]- and [9]phenacenes: (a) Full monolayer coverage of densely packed [7]phenacene. (b) Sub-monolayer coverage of [9] phenacene with the bare Ag(111) surface appearing between the molecules. (c) Overview scan of [9]phenacene. In areas with a stable adsorption geometry, molecules are resolved with submolecular resolution. Reproduced with permission from Ref. [39], Copyright 2017, American Chemical Society.

Although the Mallory photocyclization is a powerful phenacene-synthesis method, the reaction efficiency is not always very high. In order to prepare phenacenes more conveniently, a continuous flow reaction technique was applied to the Mallory photocyclization [40–43]. Continuous flow photolysis is generally advantageous over conventional batch photolysis, e.g., short irradiation time, suppression of overreaction, and no limit of reaction scale [44, 45].

Table 10.2 provides some examples for phenacene synthesis by continuous flow photolysis. A simple flow photoreactor consisting a reaction coil made of FEP (fluorinated ethylene–propylene copolymer), a high-pressure mercury arc lamp, and a Pyrex water jacket was used [46]. [n]Phenacenes (n = 4–7) were efficiently obtained in high chemical yields (73–91%) within a short irradiation time (6–10 min). The production efficiency shown in Table 10.2 means the yield of phenacenes expected for 1 h operation of the flow synthesis.

Extremely large phenacenes have been synthesized by double Mallory photocyclization protocol; thus, phenacene frameworks with benzene rings up to n = 14 were constructed [19, 37, 38, 40, 47]. Mallory reported synthesis and characterization of [11]phenacene derivative (Fig. 10.13a). The precursor, in which three phenanthrene moieties are connected with two ethylene linkers, was irradiated to afford the [11]phenacene derivative having four pentyl groups. In the electronic absorption spectrum of the substituted [11]phenacene, an absorption band assigned to the p band was observed at 400 nm. The absorption wavelength red-shifted compared to that expected from extrapolation of non-substituted lower phenacenenes. Thus, steric effects of the alkyl substituents at the "bay region" affected the electronic character to show the red-shift of the *p* band [19].

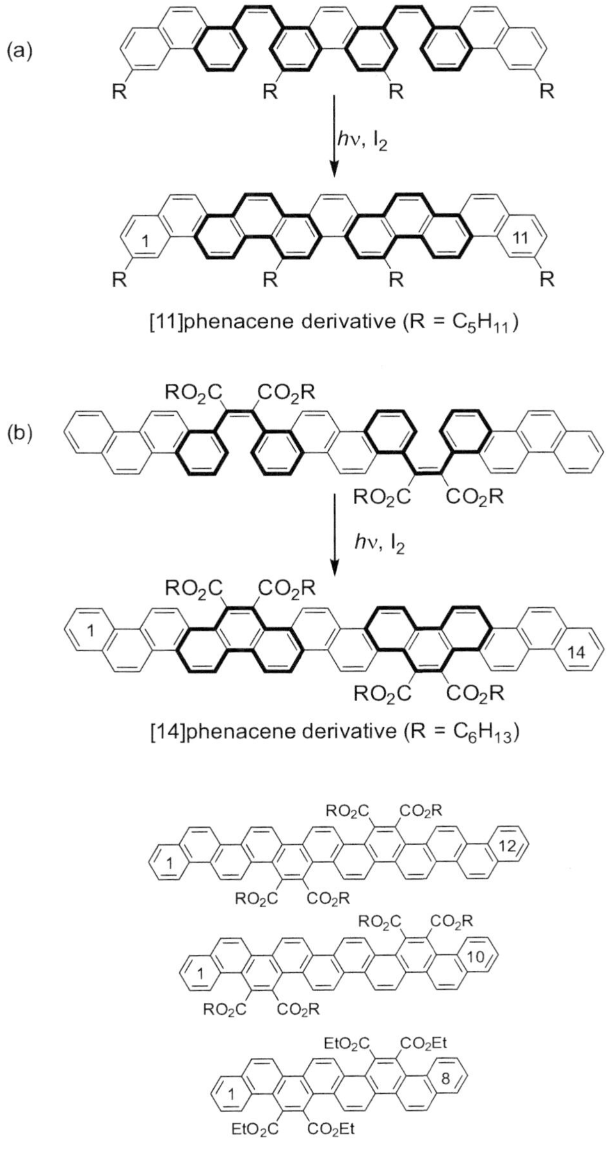

Figure 10.13 Extremely π-extended phenacenes by double Mallory photocyclization.

Table 10.2 Phenacene synthesis by continuous flow photolysis

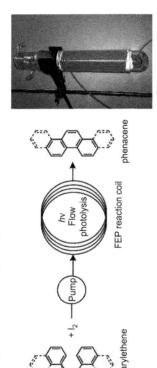

	Diarylethene substrate		Phenacene product		Residence time (Irradiation time)	Isolation yield (production efficiency)
1.		(100 mg)		(91 mg)	6 min	91% (273 mg/h)
2.		(50 mg)		(45 mg)	6 min	92% (270 mg/h)
3.		(70 mg)		(54 mg)	8 min	78% (220 mg/h)
4.		(52 mg)		(38 mg)	6 min	73% (114 mg/h)
5.		(35 mg)		(28 mg)	10 min	80% (168 mg/h)

Data taken from Ref. [42].

The double Mallory photocyclization was also applied to a substrate having three chrysene moieties, which were connected with maleic acid ester bridges (Fig. 10.13b) [47]. This reaction proceeded smoothly to produce a [14]phenacene derivative. The [14]phenacene is the largest phenacene framework experimentally obtained to date. By the same strategy, a series of substituted phenacenes of n = 8, 10, 12 were synthesized. Interestingly, the optical band gaps of the [8]~[14]phenacenes, estimated from the electronic absorption spectra, correspond to 2.99 eV for the [8]- and [10]phenacenes, 2.88 eV for the [12]- and [14]phenacenes; thus, these long phenacenes maintain very large HOMO–LUMO energy gap though the π-conjugation is extremely extended.

10.4 OFETs by Using Phenacenes as Active Layer

In the last decades, a number of PAHs and heterocyclic aromatics have been applied to the active layer of OFET devices [1–4]. Needless to say, pentacene is one of the most extensively investigated semiconductor materials. Heterocyclic aromatic systems including heteroatoms, namely, sulfur atoms, are also an important class of materials in OFET applications, and various new compounds and molecular designs are being continuously proposed [48]. As for pure aromatic hydrocarbons containing solely C and H atoms, rubrene is the benchmark compound of high-performance OFET material that displays maximum charge mobilities (μ) of 18 cm^2/Vs and 40 cm^2/Vs for two- and four-terminal single-crystal OFET devices, respectively [49].

As described in Sec. 10.2, phenacenes are very stable compounds with wide band gaps. Although fewer studies have been carried out on the applications of phenacenes, it has been revealed that phenacenes can be promising materials for the active layer of high-performance OFET devices [50]. Figure 10.14 shows the first picene-based thin-film OFET device [31]. The top contact device displayed p-channel operation under vacuum and oxygen atmospheres. The hole mobility in vacuum was estimated to be μ = 0.11 cm^2/Vs (|V_{TH}| = 37 V) for the forward transfer curve (Fig. 10.14c). The mobility increased up to μ = 1.1 cm^2/Vs after exposure to oxygen (Fig. 10.14d).

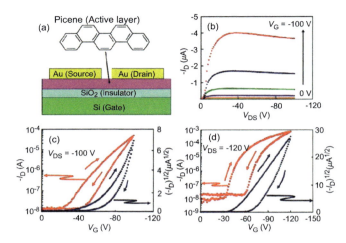

Figure 10.14 OFET device using picene thin film as the active layer. (a) The device structure, (b) output curves (vacuum), (c) transfer curve in vacuum, and (d) transfer curve after exposure to O_2. Reproduced with permission from Ref. [31], Copyright 2008, American Chemical Society.

Thin-film OFET devices have been fabricated with a series of [n]phenacenes (n = 5–9) and their performances have been systematically evaluated. The results are summarized in Table 10.3. The phenacene-based thin-film OFET devices indicated high mobility irrespective of the benzene ring number. The μ value of the [6]phenacene (fulminene)-based OFET device was the highest among those investigated.

Table 10.3 Maximum mobility of phenacene thin-film OFETs

	μ (cm^2/Vs)	$|V_{TH}|$ (V)	on/off ratio	Ref.
Picene[a]	1.4	–	–	[51]
Fulminene[a]	7.4	69	4×10^7	[52]
[7]phenacene	0.75	–	–	[53]
[8]phenacene	1.7	39	2×10^7	[28]
[9]phenacene	1.69	42	2×10^8	[23]

[a]Picene and fulminene, respectively, refer to [5]- and [6]phenacenes.

OFETs using organic single crystals have attracted much attention because they provide OFET devices that are not affected by defects, grain boundaries, and impurities, and reflect the intrinsic nature of the organic molecules. Therefore, single crystals of phenacenes have been applied to OFET devices to systematically evaluate their performances as semiconductor materials.

Figure 10.15 shows a single-crystal OFET device using [9]phenacene as the active layer. The single crystals of [9]phenacene were made by the physical vapor transport (PVT) method [23]. The device displayed clear p-channel operation with the highest μ value of 10.5 cm^2/Vs ($|V_{TH}|$ = 16.5 V, on/off ratio 5.3 × 10^8). The mobility was much higher than that observed for the thin-film device (cf. Table 10.3). Therefore, [9]phenacene single-crystal OFET displayed very high performance due to the highly ordered structure in the single crystal compared to the case of thin films. Through optimization of the [9]phenacene single-crystal OFET device configuration, the maximum μ value reached as high as 18 cm^2/Vs and a very-low-voltage operating device was attained ($|V_{TH}|$ = 2.2 V) [23].

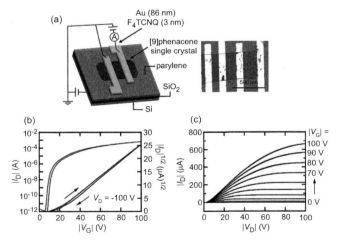

Figure 10.15 (a) Schematic diagram of single-crystal OFET of [9]phenacene with SiO$_2$ gate dielectric (left) and a picture of the [9]phenacene single-crystal (right). (b) Transfer and (c) output curves of the [9]phenacene single-crystal OFET device.

Table 10.4 gives the maximum μ values reported for single-crystal phenacene OFET devices (SiO$_2$ gate dielectric). The mobility tends to increase as the phenacene conjugation elongates. In Fig. 10.16, the μ values are plotted against the benzene ring number [n] together with those reported for single-crystal acene-based OFETs. Interestingly, both acene and phenacene series indicate quite similar tendency that charge mobility appreciably increases as a function of [n]. Therefore, extension of π-conjugation in PAHs is considered to be a promising strategy to enhance charge mobility from the aspect of molecular design.

Table 10.4 Maximum mobility of phenacene single-crystal OFETs (SiO$_2$ gate dielectric)

	μ (cm^2/Vs)	$|V_{TH}|$ (V)	on/off ratio	Ref.
Picene[a]	1.3	30	–	[54]
Fulminene[a]	0.56	36	2 × 10^9	[27]
[7]phenacene	6.9	50	1 × 10^9	[55]
[8]phenacene	8.2	26	2 × 10^8	[28]
[9]phenacene	10.5	47	2 × 10^8	[23]

[a]Picene and fulminene, respectively, refer to [5]- and [6]phenacenes.

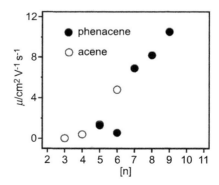

Figure 10.16 Plots of μ values reported for single-crystal phenacene OFETs of acenes and phenacenes as a function of benzene ring number [n]. Data for phenacenes are taken from references cited in Table 10.3, and those for acenes are taken from Refs. [56–59].

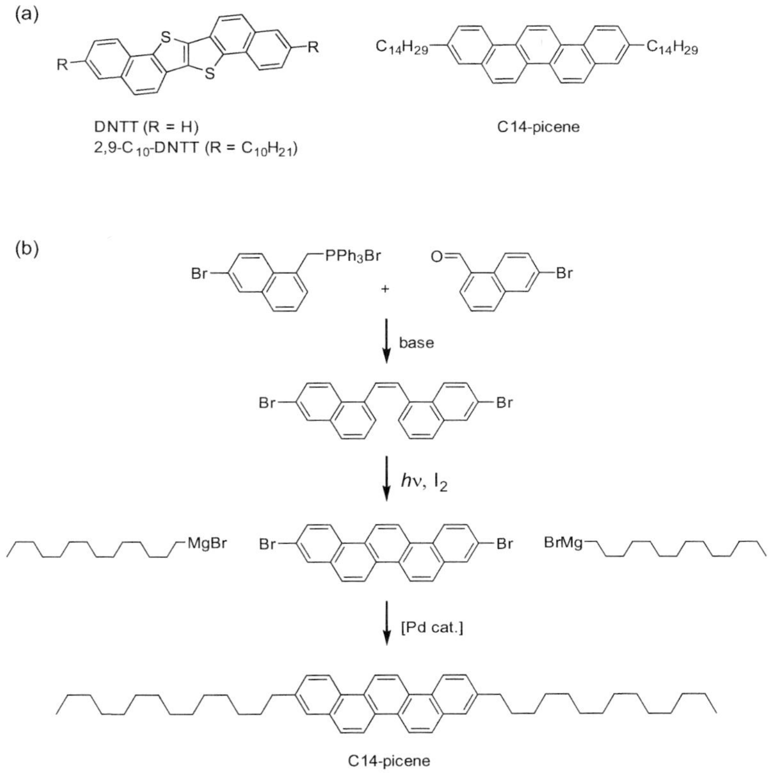

Figure 10.17 (a) Chemical structures of DNTTs and C14-picene. (b) Synthetic route to C14-picene.

Introducing long alkyl chains into OFET materials is a well-established strategy to improve charge mobility of devices. The long alkyl chains bring about tight packing of the core aromatics in the solid state due to van der Waals interactions between them. This is called "molecular fastener" or "zipper effects." The modified solid structures tend to affect OFET performance compared to the parent aromatics [48, 60]. For example, thin-film OFETs of DNTT (dinaphtho[2,3-b:2′,3′-f]thieno[3,2-b]thiophene) showed μ value as high as 3 cm^2/Vs, while decyl-substituted 2,9-C$_{10}$-DNTT showed much enhanced μ values up to 8 cm^2/Vs [48]. (See Fig. 10.17a for structures of DNTTs.) This strategy has been adopted to a picene-based OFET; thus, tetradecyl-substituted picene (C14-picene, see Fig. 10.17a for the structure of C14-picene) was conveniently synthesized via a three-step reaction sequence (Fig. 10.17b) [61]. Thin-film OFET devices have been prepared with C14-picene using various gate dielectrics. Figure 10.18 illustrates the typical OFET device structure, transfer and output curves for the C14-picene thin-film device. Table 10.5 gives the average as well as maximum μ values for the C14-picene thin-film devices. The C14-picene OFETs displayed very high mobility irrespective of the gate dielectric investigated (5–14 cm^2/Vs). It is worth noting that the C14-picene OFET device using PZT (PbZr$_{0.52}$Ti$_{0.48}$O$_3$) as the gate dielectric showed the maximum mobility as high as 20.9 cm^2/Vs, which is the highest class μ value for thin-film OFET devices. As μ values for parent picene are ~1 cm^2/Vs [54], the zipper effects of the long alkyl chains are effective to enhance phenacene-based OFET device performance.

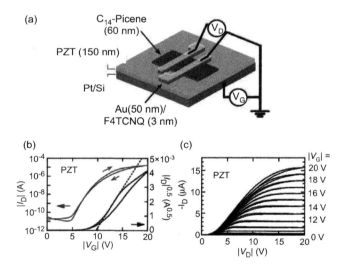

Figure 10.18 (a) Schematic diagram of thin-film OFET of C14-picene with an SiO$_2$ gate dielectric. (b) Transfer and (c) output curves of the C14-picene thin-film OFET device.

Table 10.5 The mobility of C14-picene thin-film OFET devices[a]

Gate dielectrics	μ (cm^2/Vs)[b]	$|V_{TH}|$ (V)	On/off ratio
SiO$_2$	7 (9.5)	30	6×10^6
HfO$_2$	5 (7.7)	10	5×10^5
PZT[c]	14 (20.9)	7	2×10^6
Ta$_2$O$_5$	5 (10.5)	11	3×10^5
ZrO$_2$	9 (12.0)	8	6×10^6

[a]Data taken from Ref. [61].
[b]Average values. Maximum μ values are denoted in the parentheses.
[c]PTZ:PbZr$_{0.52}$Ti$_{0.48}$O$_3$.

10.5 Concluding Remark

In this chapter, we highlighted the physical properties, efficient synthesis, and transistor applications of phenacenes. Because phenacenes are easily prepared by Mallory photocyclization and are quite chemically stable molecules, and as described in Sec. 10.4, they serve as the active layer of very high-performance OFET devices. Therefore, phenacenes are considered to be promising molecules for application in organic electronics. Still less attention has been paid to materials and applied sciences of phenacenes, including organic electronics, compared to those on acenes and heteroaromatics; however, the unique structural and electronic features of phenacenes would open avenues to new fundamental and applied sciences in the future. The discovery of novel superconductors with alkali–metal-doped picene is one of such possibilities [11].

References

1. Operamolla, G. and Farinola, G. M. (2011) *Eur. J. Org. Chem.*, 423.
2. Dong, H., Fu, X., Liu, J., Wang, Z., and Hu, W. (2013) *Adv. Mater.*, **25**, 6158.
3. Zhou, K., Dong, H., Zhang, H.-L., and Hu, W. (2014) *Phys. Chem. Chem. Phys.*, **16**, 22448.

4. Anthony, J. E. (2008) *Angew. Chem. Int. Ed.*, **47**, 452.
5. Anthony, J. E. (2006) *Chem. Rev.*, **106**, 5028.
6. Yang, X., Xu, X., and Zhou, G. J. (2015) *Mater. Chem. C*, **3**, 913.
7. Günes, S., Neugebauer, H., and Sariciftci, N. S. (2007) *Chem. Rev.*, **107**, 1324.
8. Liang, Y., Xu, Z., Xia, J., Tsai, S.-T., Wu, Y., Li, G., Ray, C., and Yu, L. (2010) *Adv. Mater.*, **22**, E135.
9. Peet, J., Heeger, A. J., and Bazan, G. C. (2009) *Acc. Chem. Res.*, **42**, 1700.
10. Lin, Y.-Y., Gundlach, D. J., Nelson, S. F., and Jackson, T. N. (1997) *IEEE Electron Device Lett.*, **18**, 606.
11. Mitsuhashi, R., Suzuki, Y., Yamanari, Y., Mitamura, H., Kambe, T., Ikeda, N., Okamoto, H., Fujiwara, A., Yamaji, M., Kawasaki, N., Maniwa, N., and Kubozono, Y. (2010) *Nature*, **464**, 76.
12. Kubozono, Y., Mitamura, H., Lee, X., He, X., Yamanari, Y., Takahashi, Y., Suzuki, Y., Kaji, Y., Eguchi, R., Akaike, K., Kambe, T., Okamoto, H., Fujiwara, A., Kato, T., Kosugi, T., and Aoki, H. (2011) *Phys. Chem. Chem. Phys.*, **13**, 16476.
13. Kubozono, Y., Eguchi, R., Goto, H., Hamao, S., Kambe, T., Terao, T., Nishiyama, S., Zheng, L., Miao, X., and Okamoto, H. (2016) *J. Phys. Condens. Matter.*, **28**, 334001.
14. Solà, M. (2013) *Front. Chem.*, **1**, 22.
15. Clar, E. (1972) *The Aromatic Sextet* (Wiley, New York).
16. Thorley, K. J. and Anthony, J. E. (2014) *Isr. J. Chem.*, **54**, 642.
17. Klessinger, M. and Michl, J. (1995) *Excited States and Photochemistry of Organic Molecules* (VCH Publisher, New York).
18. Hesse, M., Meier, H., and Zeeh, B. (2007) *Spectroscopic Methods in Organic Chemistry*, 2nd edn (Thieme, Verlag KG, Stuttgart).
19. Mallory, F. B., Butler, K. E., Evans, A. C., Brondyke, E. J., Mallory, C. W., Yang, C., and Ellenstein, A. (1997) *J. Am. Chem. Soc.*, **119**, 2119.
20. Okamoto, H., Yamaji, M., Gohda, S., Sato, K., Sugino, H., and Satake, K. (2013) *Res. Chem. Intermed.*, **39**, 147.
21. Tönshoff, C. and Bettinger, H. F. (2010) *Angew. Chem. Int. Ed.*, **49**, 4125.
22. Yasuda, T., Goto, T., Fujita, K., and Tsutsui, T. (2004) *Appl. Phys. Lett.*, **85**, 2098.
23. Shimo, Y., Mikami, T., Hamao, S., Goto, H., Okamoto, H., Eguchi, R., Gohda, S., Hayashi, Y., and Kubozono, Y. (2016) *Sci. Rep.*, **6**, 21008.

24. Mahns, B., Kataeva, O., Islamov, D., Hampel, S., Steckel, F., Hess, C., Knupfer, M., B. Büchner, Himcinschi, C., Hahn, T., Renger, R., and Kortus, J. (2014) *Cryst. Growth Des.*, **14**, 1338.
25. Holmes, D., Kumaraswamy, S., Matzger, A. J., and Vollhardt, K. P. C. (1999) *Chem. Eur. J.*, **5**, 3399.
26. Mattheus, C. C., Dros, A. B., Baas, J., Meetsma, A., de Boer, J. L., and Palstra, T. T. M. (2001) *Acta Cryst.*, **C57**, 939.
27. He, X., Eguchi, R., Goto, H., Uesugi, E., Hamao, S., Takabayashi, Y., and Kubozono, Y. (2013) *Org. Electron.*, **14**, 1673.
28. Okamoto, H., Eguchi, R., Hamao, S., Goto, H., Gotoh, K., Sakai, Y., Izumi, M., Takaguchi, Y., Gohda, S., and Kubozono, Y. (2014) *Sci. Rep.*, **4**, 5330.
29. Burg, O. (1880) *Ber. Dtsch. Chem. Ges.*, **13**, 1834.
30. Bamberger, E. and Chattaway, F. D. (1895) *Liebigs Ann.*, **284**, 52.
31. Okamoto, H., Kawasaki, N., Kaji, Y., Kubozono, Y., Fujiwara, A., and Yamaji, M. (2008) *J. Am. Chem. Soc.*, **130**, 10470.
32. Harvey, R. G., Pataki, J., Cortez, C., Di Raddo, P., and Yang, C. (1991) *J. Org. Chem.*, **56**, 1210.
33. Okamoto, H., Yamaji, M., Gohda, S., Kubozono, Y., Komura, N., Sato, K., Sugino, H., and Satake, K. (2011) *Org. Lett.*, **13**, 2758.
34. Carrera, M., Viuda, M., and Guijarro, A. (2016) *Synlett*, **27**, 2783.
35. Jørgensen, K. B. (2010) *Molecules*, **15**, 4334.
36. Mallory, F. B. and Mallory, C. W. (1984) *Org. React.*, **30**, 1.
37. Mallory, F. B., Butler, K. E., Evans, A. C., and Mallory, C. W. (1996) *Tetrahedron Lett.*, **37**, 7173.
38. Mallory, F. B., Butler, K. E., Bérubé, A., Luzik, E. D., Mallory, C. W., Brondyke, E. J., Hiremath, R., Ngo, P., and Carroll, P. J. (2001) *Tetrahedron*, **57**, 3715.
39. Chen, S.-W., Sang, I.-C., Okamoto, H., and Hoffmann, G. (2017) *J. Phys. Chem. C*, **121**, 11390.
40. Hernandez-Perez, A. C., Vlassova, A., and Collins, S. K. (2012) *Org. Lett.*, **14**, 2988.
41. Lefebvre, Q., Jentsch, M., and Rueping, M. (2013) *Beilstein J. Org. Chem.*, **9**, 1883.
42. Okamoto, H., Takane, T., Gohda, S., Kubozono, Y., Sato, K., Yamaji, M., and Satake, K. (2014) *Chem. Lett.*, **43**, 994.
43. Okamoto, H., Takahashi, H., Takane, T., Nishiyama, Y., Kakiuchi, K., Gohda, S., and Yamaji, M. (2017) *Synthesis*, **49**, 2949.

44. Knowles, J. P., Elliott, L. D., and Booker-Milburn, K. I. (2012) *Beilstein J. Org. Chem.*, **8**, 2025.
45. Mizuno, K., Nishiyama, Y., Ogaki, T., Terao, K., Ikeda, H., and Kakiuchi, K. (2016) *J. Photochem. Photobiol. C*, **29**, 107.
46. Hook, B. D. A., Dohle, W., Hirst, P. R., Pickworth, M., Berry, M. B., and Booker-Milburn, K. I. *J.* (2005) *Org. Chem.*, **70**, 7558.
47. Moreira, T. S., Ferreira, M., Dall'armellina, A., Cristiano, R., Gallardo, H., Hillard, E. A., Bock, H., and Durola, F. (2017) *Eur. J. Org. Chem.*, 4548.
48. Takimiya, K., Osaka, I., Mori, T., and Nakano, M. Acc. (2014) *Chem. Res.*, **47**, 1493.
49. Takeya, J., Yamagishi, M., Tominari, Y., Hirahara, R., Nakazawa, Y., Nishikawa, T., Kawase, T., Shimoda, T., and Ogawa, S. (2007) *Appl. Phys. Lett.*, **90**, 102120.
50. Kubozono, Y., He, X., Hamao, S., Teranishi, K., Goto, H., Eguchi, R., Kambe, T., Gohda, S., and Nishihara, Y. (2014) *Eur. J. Inorg. Chem.*, **24**, 3806.
51. Kawasaki, N., Kubozono, Y., Okamoto, H., Fujiwara, A., and Yamaji, M. (2009) *Appl. Phys. Lett.*, **94**, 043310.
52. Eguchi, R., He, X., Hamao, S., Goto, H., Okamoto, H., Gohda, S., Sato, K., and Kubozono, Y. (2013) *Phys. Chem. Chem. Phys.*, **15**, 20611.
53. Sugawara, Y., Kaji, Y., Ogawa, K., Eguchi, R., Oikawa, S., Gohda, H., Fujiwara, A., and Kubozono, Y. (2011) *Appl. Phys. Lett.*, **98**, 013303.
54. Kawai, N., Eguchi, R., Goto, H., Akaike, K., Kaji, Y., Kambe, T., Fujiwara, A., and Kubozono, Y. (2012) *J. Phys. Chem. C*, **116**, 7983.
55. He, X., Hamao, S., Eguchi, R., Goto, H., Yoshida, Y., Saito, G., and Kubozono, Y. (2014) *J. Phys. Chem. C*, **118**, 5284.
56. Aleshin, A. N., Lee, J. Y., Chu, S. W., Kim, J. S., and Park, Y. W. (2004) *Appl. Phys. Lett.*, **84**, 5383.
57. de Boer, R. W. I., Klapwijk, T. M., and Morpurgo, A. F. (2003) *Appl. Phys. Lett.*, **83**, 4345.
58. Goldmann, C., Haas, S., Krellner, C., Pernstich, K. P., Gundlach, D. J., and Batlogg, B. (2004) *J. Appl. Phys.*, **96**, 2080.
59. Watanabe, M., Chang, Y. J., Liu, S.-W., Chao, T.-H., Goto, K., Islam, M. M., Yuan, C.-H., Tao, Y.-T., Shinmyozu, T., and Chow, T. J. (2012) *Nat. Chem.*, **4**, 574.
60. Inokuchi, H., Saito, G., Wu, P., Seki, K., Tang, T. B., Mori, T., Imaeda, K., Enoki, T., Higuchi, Y., Inaka, K., and Yasuoka, N. (1986) *Chem. Lett.*, **15**, 1263.
61. Okamoto, H., Hamao, S., Goto, H., Sakai, Y., Izumi, M., Gohda, S., Kubozono, Y., and Eguchi, R. (2014) *Sci. Rep.*, **4**, 5048.

Index

acceptor 29, 67, 82, 101, 156, 160, 162, 163, 166, 167
acene 173–178, 180–182, 184, 186, 188, 190, 196–200, 212, 215
acid
 barbituric 77, 80
 citric 17
 croconic 126, 127
 cyanuric 77
 nitric 4
 trifluoroacetic 4
active layer 94, 177, 181, 187, 189, 190, 209–211, 213, 215
AFM *see* atomic force microscopy
aggregates 164, 180, 181
 ill-defined 66
 irregular 61
aggregation-induced delayed fluorescence 164
aggregation-induced emission (AIE) 30, 164
AIE *see* aggregation-induced emission
alkyl chain 6, 78–81, 86, 100, 213, 214
 branched 10, 11, 13, 16, 21
 hydrophobic 18
anthracene 7, 8, 44, 46, 139, 141, 181, 198, 199
aromatic hydrocarbon 7, 173, 209
aryl gold isocyanide complex 136, 137
atomic force microscopy (AFM) 85, 179, 185
azobenzene 11, 28, 40–42

BADGE *see* bisphenol A diglycidyl ether
Baird aromaticity 29
band
 photoabsorption 67
 Soret 116
Beer–Lambert law 105
benzophenone 155
BHJ *see* bulk heterojunction
BHJ layer 95, 103, 181, 189
BHJ solar cell 76, 83, 85, 86, 88, 89
bias 102–104, 106, 107, 109
bisphenol A diglycidyl ether (BADGE) 38, 39
BODIPY *see* boron-dipyrromethene
Boltzmann constant 105
boron-dipyrromethene (BODIPY) 34, 67, 69
bulk heterojunction (BHJ) 76, 94, 99, 181, 188, 189

carbazole 3, 157, 160
carbazolyl dicyanobenzene 157
carbon quantum dot 17
cavity 20, 58, 96, 97, 103, 122, 126
 central 122
 empty 103
 resonant 96, 97
charge recombination (CR) 94, 95, 101, 107, 108, 153
charge separation (CS) 76, 78, 94, 95, 98, 99, 101, 102, 108, 122
charge transfer (CT) 30, 94, 156
charge transfer
 bond-cleavage-induced intramolecular 34
 twisted intramolecular 28
CHL *see* cyclic host liquid

circularly polarized luminescence 167
columnar nanostructure 76, 77, 83
column chromatography 119, 124, 125
compound
  aromatic 202
  benchmark 173
  liquid crystalline 10
  low-melting 3
  mechano-responsive 135
  organometallic 134
  π-conjugated 15
  photochromic 28, 29
  polycyclic aromatic 196
conjugated polymer microresonator 58, 59, 71
conjugated polymer microsphere 57, 59, 60, 63–65, 69, 71, 99, 101, 105
contact film transfer method 98
COT *see* cyclooctatetraene
CR *see* charge recombination
crystallinity 60–62, 69, 71, 136, 142, 177, 184–186
crystallization 3, 5, 141, 178
CS *see* charge separation
CT *see* charge transfer
cyanovinyljulolidine 34
cychlopentadithiophene 98
cyclic host liquid (CHL) 20, 21
cyclo[8]isoindole 117, 118
cyclodextrin 41, 42
cyclooctatetraene (COT) 29–31, 34, 35, 44, 45

delayed fluorescence 155, 157–159, 162, 164, 166
density functional theory 61
derivative
  carborane 164
  fulleropyrrolidine 6
  porphyrin 155

device 85, 102–104, 151, 178, 187–189, 209, 211, 213
  as-cast 85
  foldable 2
  layered 102, 103
  memory 144, 165
  photofunctional 128
  semiconductor 174
  solar cell 84, 85, 109
  thin-film 211, 214
dicyanobenzene 157
dicyanoquinoxaline 160
Diels–Alder activity 42
diffusion theory 100, 109
donor 29, 76, 82, 94, 101, 102, 156, 158–162, 167

electric field 94, 96, 103, 105, 109
electric-field-induced birefringence 12
electric-field-induced second harmonic 102
electroluminescence 151
emission 8, 9, 17, 30, 35, 36, 69, 135, 139–144, 153, 157–159, 165
  dual 166
  excimer 9
  orange 139, 157, 166
  red 134
  short wavelength 142
  white 166
  yellow 166
emission layer 65, 152, 153, 164
emission mechanism 140–143, 154
energy
  binding 4
  electron exchange 155
  electron repulsion 155
  photon 95, 96
  rotational 96
energy transfer 59, 67, 68, 162
  cavity-mediated 67

fluorescence resonance 30
EQE *see* external quantum efficiency
ESIPT *see* excited-state intramolecular proton transfer
ethylene glycol diglycidyl ether 39
exciplex 162, 163
excitation 32, 33, 48, 50, 63, 65, 69, 96, 99, 106, 140, 175
  electrical 153
  electronic 95
  laser 64, 65, 68
excitation wavelength 30, 66, 100
excited state 29, 30, 49, 68, 141, 153, 154, 156, 157, 162
excited-state intramolecular proton transfer (ESIPT) 30, 40
exciton 94, 152, 153, 186
exciton diffusion 99, 101
external quantum efficiency (EQE) 85, 86, 101

FET *see* field-effect transistor
field-effect transistor (FET) 1, 98, 102
film
  adhesive 46–48, 50
  as-cast 82, 84
  blend 19, 82, 83, 181, 186
  cast 66
  drop-cast 18
  low-crystallinity 179
  thermally annealed 82
  thin 45, 86, 87, 152, 175–177, 179–181, 184, 186, 190, 191, 210, 211
  vacuum-deposited 177, 179, 181
  wet 185
FLAP *see* flexible and aromatic photofunctional system
FLAP1 30–32, 39
FLAP2 34–39
FLAP3 35, 36

FLAP4 34, 37, 38
FLAP5 44–50
FLAP6 45
FLAP7 50
flexible and aromatic photofunctional system (FLAP) 30, 31, 37–40, 45, 51
fluorescence 33, 36, 50, 140, 141, 154
  excimeric 10
  green 30, 49, 50
fluorescent dye 58, 59, 67, 71
fluorophore 30, 33
FML *see* functional molecular liquid
Förster resonance energy transfer (FRET) 8, 10, 30
FRET *see* Förster resonance energy transfer
fullerene 5, 7, 94, 155
fullerene heterojunction 93, 98
fulminene 197, 199, 200, 202, 203, 210, 212
functional molecular liquid (FML) 2–5, 7, 19, 21

glassy transition 6, 10
gold isocyanide 135, 137, 140, 144
Guerbet alcohol 5

highly oriented pyrolytic graphite (HOPG) 78, 79, 86, 87
HOPG *see* highly oriented pyrolytic graphite
Hückel antiaromaticity 29
Hückel aromaticity 116

insulated molecular wire 18

light pulse 95, 96, 103
lowest unoccupied molecular orbital (LUMO) 118, 142, 156, 158–162, 164, 167, 199, 201
luminescence 9, 30, 144

luminescent mechanochromism 133, 135–137, 139, 141, 144
LUMO *see* lowest unoccupied molecular orbital

Mallory photocyclization 203–206, 215
mechanical stimulation 133–138, 140, 141, 143, 144
mechanochromic compound 133–135, 138, 139, 141, 144
microcavity 65, 66
microwave 94–96, 102, 103
miniemulsion method 59, 60, 62, 63, 69
Möbius aromaticity 29, 116
molecular rotor 33, 34, 37, 38
molecule
  asymmetric 164
  cage 20
  decomposed 8
  discoid 76
  electron-acceptor 162
  emitter 151, 152
  excited 50
  fluorescent 39
  organic 1, 151, 154, 196, 211
  π-conjugated 29
  polar 96
  triphenylborane-based 161

Newtonian fluid 33, 37
nonacene 175, 176

OEG *see* oligo(ethylene glycol)
OFET *see* organic field-effect transistor
OLED *see* organic light-emitting diode
oligo(ethylene glycol) (OEG) 2–4
optical microresonator 57, 58
organic electronics 58, 191, 196, 215

organic field-effect transistor (OFET) 85, 177, 182, 190, 195, 209, 211, 213, 214
organic laser 151
organic light-emitting diode (OLED) 8, 12, 151–155, 157, 163–165, 167, 196
organic solar cell (OSC) 174, 176, 178, 180, 182, 184, 186, 188, 190
OSC *see* organic solar cell
oxidative coupling 118, 120–124, 126–128

PAH *see* polycyclic aromatic hydrocarbon
PCBM 83, 98, 101, 104–109, 190
$PC_{61}BM$ 78, 81–86, 181
$PC_{71}BM$ 82, 88, 101, 104–108, 181–186, 188, 189
PCE *see* power conversion efficiency
π-conjugated polymer 18, 27, 57–62, 64, 66, 68, 70–72, 95, 97
PCPDTBT 61, 98–101, 104–109
PDMS *see* poly(dimethylsiloxane)
pentacene 173–179, 196–201, 209
pentaphyrin 115, 116
phenacene 196–215
phenanthrene 196, 198, 199, 204
phosphorescence 30, 140, 141, 153–155
photodimerization 41, 45, 46, 49
photoexcitation 32, 45, 49, 57, 69, 153
photoirradiation 41, 65, 66, 137, 176, 188
photoisomerization 11, 12, 34
photoluminescence 1, 15, 57, 136, 138, 141, 143, 144, 152
photoprecursor approach 176–189

photoreaction 45, 177, 178, 181, 182, 185, 203
phthalocyanine 3, 15, 16
physical vapor transport 211
picene 196–200, 202, 203, 210, 212–214
  metal-doped 215
  tetradecyl-substituted 214
polarized optical microscopy (POM) 10, 45, 122, 123
poly(dimethylsiloxane) (PDMS) 2, 4, 5
polycyclic aromatic hydrocarbon (PAH) 195, 197, 212
polyfluorene 18, 61
polymer 1, 31, 42, 58–63, 66, 69, 94, 96, 98–100, 102, 104–106, 108
  amorphous 104
  carbazole 65
  diarylethene-containing 41
  face-on-rich 108
  low band-gap 98
  orientation-controlled 93
  silicon-based organic 4
  supramolecular 77
polyphenylenevinylene (PPV) 18, 61, 65, 69, 70
POM *see* polarized optical microscopy
porphyrin 15, 16, 29, 115–117, 122
powder X-ray diffraction (PXRD) 77, 142
power conversion efficiency (PCE) 78, 82–84, 88, 93, 94, 102, 108, 183, 186, 189
PPV *see* polyphenylenevinylene
PXRD *see* powder X-ray diffraction
pyrene 3, 7, 9

Raman spectroscopy 95
relaxation 105, 107, 108, 141
resonator 58, 63

reverse intersystem crossing (RISC) 153–156, 162, 163
ring-expanded porphyrins 115, 116, 118, 120, 122, 124, 126, 128
RISC *see* reverse intersystem crossing
rosette 76, 77, 79, 80, 83, 85, 87, 88
  barbituric acid 77
  enantiomeric 87
  heptameric 85
  hexameric 77
  hydrogen-bonded 78, 79
  six-membered 80
rotaxane 20, 21
rubrene 12, 13, 209

sapphyrin 115, 116
scanning tunneling microscopy (STM) 80, 81, 85, 205
SCLC *see* space-charge-limited current
self-assembled microsphere 57, 59, 61, 69, 70, 95, 97
self-organization precipitation method 60
sextetthiophene 83
solar cells 1, 58, 94, 102, 108
  organic 175, 181, 182
space-charge-limited current (SCLC) 102, 104, 179
STM *see* scanning tunneling microscopy
Stokes shift 16, 35, 36
Suzuki–Miyaura cross-coupling 125

TADF *see* thermally activated delayed fluorescence
TADF emitter 152, 155, 157, 168
TADF material 152, 154, 156, 168
TADF-OLED 152, 153, 155, 157–161, 164, 167, 168

TAS *see* transient absorption spectroscopy
tetrahydrofuran (THF) 35, 37, 59, 60
thermal annealing 18, 82, 84, 85, 88, 166, 183
thermally activated delayed fluorescence (TADF) 30, 152–156, 158–160, 162, 164, 166–168
THF *see* tetrahydrofuran
thiophene 126, 163, 213
TICT *see* twisted intramolecular charge transfer
time-resolved microwave conductivity (TRMC) 10, 78, 93, 94, 98, 99, 102, 103, 107, 109
TPC *see* transient photocurrent
transient absorption spectroscopy (TAS) 101, 106
transient photocurrent (TPC) 102, 104–107, 109
transition 4, 13, 141, 142, 154, 161, 199
transparent conductive oxide 103
TRMC *see* time-resolved microwave conductivity
TRMC transient 101–105, 107
twisted intramolecular charge transfer (TICT) 28, 33, 40

vapor pressure osmometry 80
viscosity 4, 6, 9, 10, 13, 14, 20, 33, 34, 37, 38

WGM *see* whispering gallery mode
WGM lasing 57, 69, 105
whispering gallery mode (WGM) 57, 63–66, 71

X-ray crystallographic analysis 119, 122, 126, 166
X-ray diffraction 98, 141, 184, 205

zipper effects 213, 214

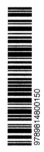